LIFE AFTER MILITARY

A COMPREHENSIVE GUIDE FOR VETERANS AND THEIR FAMILIES NAVIGATING THE TRANSITION TO CIVILIAN LIFE

MS. VETERAN AFFAIRS

STRIVE press

Dedication

To the brave men and women who have served, are serving, and those who will serve: this book is for you. May it honor your dedication, courage, and sacrifice and serve as a beacon of guidance as you navigate new paths and embrace new beginnings.

To your families, whose unwavering support and resilience make this journey possible, thank you for your strength. Your role is immeasurable, your presence irreplaceable.

May this book stand as a reminder of your boundless potential beyond service and the gratitude of a nation forever in your debt.

With respect and admiration,
Ms. Veteran Affairs

WELCOME MESSAGE

What happens when a soldier returns home, but the battle continues?

I know this all too well. My journey began in Albuquerque, New Mexico, where I was the wife of an Air Force veteran who served as an analyst for 26 years. When my husband retired, our lives took a challenging turn that we never anticipated. With four young children to care for, we had to make the difficult decision to divide our family. My husband worked in Lancaster, CA, while living in Long Beach, and I stayed in South Carolina to keep the family together, though we didn't even have a home of our own.

The challenges didn't end there. We were completely in the dark about the benefits my husband had earned through his service. When he fell ill, suffering from COPD, diabetes, and complications from Agent Orange exposure, I knew I had to step up. My husband was sick, relying on a machine for his lung condition—an ailment directly tied to his service. Navigating the complex and bureaucratic maze of veteran support was overwhelming, and the lack of guidance only made it harder.

There were times we were told my husband wasn't eligible for disability benefits, and that rejection was like a punch to the gut. But I persisted. Through education, relentless effort, and a lot of trial and error, I fought to secure the benefits we were entitled to. Eventually, I proved that my husband was eligible for 70% disability. That victory ignited a passion in me—a passion to ensure that no other veteran or their family would have to endure what we did. This drive led me to create New Horizons for Life.

This organization was born out of necessity, but it's grown into something far greater. New Horizons for Life offers benefits counseling, job employment assistance, workshops, transition support, and temporary housing for veterans. We're not just helping veterans—we're building a community. Our future goals include a community garden in Lancaster, where veterans can connect with the land and each other, and we're expanding our outreach with feeding events every 2nd and 4th Thursday, collaborating with local groups to make sure no one is left behind.

When my husband passed away, I had no time to grieve. I had to ensure my children's needs were met, finish my own education in Social and Behavioral Science, and keep pushing forward. But I never lost sight of what really matters—helping others navigate the challenges I faced.

New Horizons for Life isn't just an organization; it's a lifeline for those who resonate with my story. If only I had known then what I know now, I could have done so much more for my husband and our children. But now, with the knowledge and experience I've gained, I'm here to make sure that no other family has to go through this alone. Together, we can honor the sacrifices made and build a future that reflects the resilience and courage of those who have served.

1

UNDERSTANDING THE TRANSITION

"Courage is not the absence of fear, but the capacity to act despite our fears."

— John McCain

What happens when you hang up your uniform for the last time? When the structure and certainty of military life give way to a new, unfamiliar world? The transition from military to civilian life is one of the most profound changes a veteran can face —a journey that demands both courage and resilience.

But you don't have to face it alone. To help guide you through this journey, we introduce the S.O.A.R. framework —a structured approach designed to support you at every step. Before we dive into the specifics, let's explore what this transition really means.

The Reality of Transition

. . .

LEAVING military service isn't just about changing jobs or relocating; it's about redefining who you are and what you stand for in a civilian world that operates by different rules. It can feel like stepping into a void, where the familiar routines, sense of mission, and camaraderie are suddenly replaced by uncertainty and isolation.

You might find yourself asking, "Who am I without my uniform?" or "Where do I go from here?" These are normal questions—questions that countless veterans have faced before you. The transition involves not just practical adjustments but emotional and psychological shifts as well. It's about finding a new sense of purpose and navigating a world that may seem indifferent to the sacrifices you've made.

This is where the S.O.A.R. framework comes in—a tool to help you break down the complexities of this transition into manageable steps, so you can approach each phase with confidence and clarity.

The S.O.A.R. framework serves as a structured approach for veterans transitioning from military to civilian life. It emphasizes four key components: Self-Awareness, Opportunity, Action, and Resilience. Each component is designed to facilitate a smoother transition by encouraging veterans to reflect on their experiences and skills, explore new opportunities, take actionable steps, and develop resilience in the face of challenges.

Self-Awareness

The journey begins with Self-Awareness, where veterans are encouraged to reflect on their identities beyond military service. This involves identifying personal values, passions, and goals. Understanding one's strengths and weaknesses is crucial, as it helps veterans reconnect with their true selves and define what matters most in their new civilian lives.

Opportunity

The next step is Opportunity, which focuses on recognizing the potential career paths and educational opportunities available to veterans. The skills acquired during military service are often transferable to civilian roles, and veterans are urged to explore how their military experiences can be adapted to new contexts. This exploration can lead to fulfilling careers that align with their skills and interests.

Action

Action emphasizes the importance of taking concrete steps towards achieving goals. This can involve applying for jobs, enrolling in educational programs, or networking with others. Starting with small actions can create momentum, helping veterans move closer to their envisioned futures.

Resilience

Finally, Resilience prepares veterans for the inevitable challenges that arise during their transition. It is about learning to bounce back from setbacks and viewing challenges as opportunities for growth. Developing resilience is essential for navigating the emotional and psychological aspects of transitioning, as it fosters a mindset that embraces learning from experiences.

Emotional and Psychological Landscape

Transitioning from military to civilian life is not only a practical challenge but also an emotional and psychological one. Veterans often face feelings of loss, fear, and isolation. Acknowledging these feelings is vital, and the S.O.A.R. framework addresses both the practical and emotional aspects of this significant life change. Support systems, stress management, and community engagement can play crucial roles in this process.

By following the S.O.A.R. framework, veterans can better navigate their transition, finding fulfillment and purpose in their new civilian lives.

Mark's Journey: From Service to New Purpose with Mrs. Veteran Affairs

Let me tell you about Mark, a veteran who, like so many of you, found himself standing at the crossroads between his military service and civilian life. And I had the privilege of walking alongside him as he navigated this new chapter, full of uncertainties but also full of potential.

When Mark hung up his uniform for the last time, he stood in front of his mirror, staring at the reflection of someone he wasn't sure he recognized. For years, he'd been Sergeant Mark Davis—a leader, a protector, a soldier. Now, dressed in civilian clothes, he felt a wave of confusion wash over him. Who was he without the uniform? What purpose could he possibly serve now?

We talked about that feeling, and I reminded him that it's not just a job he's leaving behind—it's a life, a structure, a family. And it's okay to feel unsettled. The fear? That's part of it, too. Courage isn't about being fearless; it's about showing up and moving forward in spite of the fear.

One evening, after a particularly tough day, Mark took my advice. He grabbed a notebook and started writing down everything that still mattered to him—his family, loyalty, making a difference in the lives of others. That's when it clicked. The uniform may be gone, but the core of who he was hadn't changed. His values, his desire to lead and mentor—they were still there. This was Mark's first big step in the S.O.A.R. framework we use—**Self-Awareness**. He was reconnecting with the person behind the rank.

With that clarity, we started talking about how his military skills could transition into civilian opportunities. Mark didn't just see himself as a veteran anymore—he saw the wealth of experience he had in logistics, team management, and leadership. Together, we explored roles in project management, and it was like a lightbulb went off. This was a natural extension of what he'd done in service. That's the power of recognizing **Opportunity**—realizing that the tools you've honed in the military are just as valuable in the civilian world.

But recognition isn't enough. You have to take **Action**, and that's exactly what Mark did. He enrolled in a project management course, updated his resume, and started applying for jobs. I won't lie to you, the process wasn't always smooth. There were rejections, moments of self-doubt, and plenty of frustration. But with each setback, Mark got up and kept going.

This is where **Resilience** became Mark's biggest asset. It's what kept him moving when things didn't go as planned. When those rejection emails came in, we worked together to refine his approach. Each "no" wasn't a defeat—it was a learning opportunity. And slowly, he got stronger, more confident, and more determined.

Eventually, Mark landed a job that not only utilized his skills but reignited his passion for leadership and mentoring others. That sense of purpose he feared he'd lost? He found it again. And it wasn't just about the job; it was about the journey he took to get there—the growth, the perseverance, and the courage to rebuild his identity.

Transitioning from military life to civilian life is one of the hardest journeys a veteran will face. But you don't have to face it alone. There are people like me, resources like the S.O.A.R. framework, and a community of veterans who understand exactly what you're going through.

So, when you look in the mirror and wonder who you are now, remember this: The courage that led you through service is the same courage that will guide you through this transition. And with that, you won't just find your new path —you'll thrive on it, just like Mark did.

Action Steps for Your Transition Journey:

Ms. Veteran Affairs' Checklist

☐ **Reflect on Your Identity**

Take time to reflect on who you are beyond your military service. Write down your core values, passions, and goals. This will help reconnect with your true self and guide your next steps.

□ **Explore Career and Education Opportunities (Opportunity)**

Identify the skills you've developed in the military and research how they translate into civilian roles. Consider career paths or educational programs that align with your strengths and interests.

□ **Take Action Towards Your Goals (Action)**

Start by updating your resume, enrolling in relevant courses, or applying for jobs. Set small, achievable goals to build momentum and move closer to your desired future.

□ **Develop Resilience (Resilience)**

Prepare yourself for setbacks. Practice bouncing back by learning from each challenge. Maintain a positive mindset and reach out for support when needed.

□ **Engage with Your Support System (Emotional & Psychological Landscape)**

Recognize the emotional and psychological aspects of your transition. Stay connected with fellow veterans, family, or professional support systems to manage stress and feelings of isolation.

2

SELF-ASSESSMENT

"Knowing yourself is the beginning of all wisdom."

— Aristotle

Transitioning from military to civilian life is a journey filled with uncertainty, but it's also an opportunity to rediscover who you are beyond the uniform. Self-assessment is the cornerstone of this journey. It's about understanding the strengths, skills, and values that define you, both personally and professionally.

Discovering Your Strengths, Skills, and Values

Your journey starts with understanding what you bring to the table. Let's break it down:

. . .

1. **Identifying Your Strengths** Think back to your time in the service. What moments made you proud? When did you feel most effective? Your strengths might include leadership, problem-solving, or resilience under pressure. Here's a simple exercise:

○ **Strengths Inventory Exercise:** Reflect on three situations during your service where you felt particularly proud or successful. Write down the key strengths or skills you used in each situation.

▪ *Example*: If you led a team through a high-pressure mission, your strengths might include leadership, strategic thinking, and teamwork.

2. **Recognizing Your Skills** Military life equips you with a wide range of skills, both technical and soft. But do you know which ones you truly excel at or enjoy using?

○ **Skills Mapping Exercise:** Create two columns labeled "Technical Skills" and "Soft Skills." List the skills you've gained under each category. Highlight those you enjoyed most and excelled at.

▪ *Example*: If you were in logistics, you might list "supply chain management" under Technical Skills and "problem-solving" under Soft Skills.

3. **Clarifying Your Values** Your values are the compass that guides your decisions. What mattered most to you in the military? Integrity? Loyalty? These values can help you find a civilian role that aligns with your true self.

○ **Values Assessment Exercise:** Write down a list of values that were important to you during your service. Rank them in order of importance. Reflect on how these values can translate into your civilian life and career.

Reflecting on Personal and Professional Challenges

The transition isn't just about identifying strengths; it's also about recognizing the challenges you've faced and how you've overcome them. Understanding these challenges can help you anticipate and navigate the obstacles ahead.

- Challenge Reflection Exercise: List some of the biggest challenges you faced during your service. Note how you addressed them and what you learned.
  - *Example*: If you struggled with balancing work and personal life, reflect on how you managed stress and what strategies helped you cope.

Here's where the story of Maria, an Army officer, comes in. Maria faced the challenge of reintegrating with her family after long deployments. The shift from military to family life was tough, but by reflecting on her experiences, she realized that her ability to adapt and communicate under stress was a strength she could use in her transition. Today, she works as a family therapist, helping other veterans and their families navigate similar challenges.

Setting Personal Goals

Once you've assessed your strengths, skills, values, and challenges, it's time to set personal goals. Your goals should be SMART (Specific, Measurable, Achievable, Relevant, Time-bound) to keep you focused and motivated.

- Goal-Setting Worksheet:
  - **Career Goals:** What type of work are you interested in? Set specific goals related to roles or industries you want to pursue.
  - **Personal Goals:** Define what you want to achieve in your personal life, like work-life balance, hobbies, or family time.
  - **Development Goals:** Identify areas where you need further development or training, and set goals to address these needs.
    - *Example*: If you're interested in project management, a SMART goal might be: "Enroll in a project management certification course within the next three months and complete it within six months."

Moving Forward: Self-Assessment as a Lifelong Process

Self-assessment isn't a one-time task; it's an ongoing process that evolves as you do. Just like in the military, where you constantly adapted to new missions and challenges, your civilian life will require you to regularly reassess and adjust your goals.

Remember, you're not alone in this journey. Whether it's through mentorship, community, or professional support, there are resources to help you along the way. The S.O.A.R. framework is here to guide you—not just through the transition, but throughout your life as you continue to grow and adapt.

So, take that first step. Complete the exercises in this chapter, set your goals, and start your journey of self-discovery. The more you understand yourself, the better equipped you'll be to build a life that reflects who you truly are.

From Uniform to New Horizons: A Veteran's Journey to Finding Purpose After 26 Years

When my husband completed his 26-year career as an analyst in the Air Force, we all felt the weight of the transition. For more than two decades, his life had been defined by routine, mission-driven tasks, and the close-knit community of the military. But when he finally hung up his uniform, we found ourselves at a turning point, unsure of what was next.

The first few months were especially tough. After a lifetime of gathering data, making critical decisions, and supporting operations behind the scenes, civilian life seemed foreign to him. He started applying for jobs, but nothing felt right. The rejection letters piled up, and with each one, I could see the frustration building. His sense of purpose, which had always been so clear in the Air Force, now felt distant.

One night, as he sat at the dinner table, staring at yet another job application, I finally said, "You've been working nonstop for 26 years—maybe it's time to take a step back and figure out what truly excites you before jumping into something new."

It was a pivotal moment. I could see the shift in his expression, and for the first time in months, he allowed himself to pause. The next day, he grabbed a notebook and started jotting down the moments that defined his military career. It wasn't just about crunching numbers or making sense of intelligence reports—he started to remember the bigger impact he had. He had led teams during high-pressure situations, mentored younger analysts, and helped shape critical decisions. He was more than his title; he was a leader, a mentor, and someone who thrived on solving problems that mattered.

But there was something else he uncovered in those reflections: a desire for balance. After years of deployments and long hours, he wanted more time for us—for family dinners, for watching our kids grow up, and for enjoying the simple things he'd missed while serving. The clarity he gained from that self-reflection became a turning point in his transition.

He didn't rush back into the job hunt. Instead, he began researching career paths that would allow him to combine his analytical skills with his desire for balance. Project management stood out, especially in industries that prioritized service and leadership. He looked into roles in nonprofits and organizations focused on community outreach, where his leadership abilities could make a meaningful impact.

But he didn't stop at research. He took action, enrolling in project management courses to refine his skills for the civilian world. He also began volunteering with local veterans' organizations, finding ways to give back while gaining real-world experience. This period of growth wasn't easy, but it gave him confidence and a sense of purpose.

Eventually, he landed a job as a project manager at a

nonprofit that provides services to veterans. It was the perfect fit—a role that allowed him to use his analytical mind and leadership skills while staying connected to a mission-driven cause. More importantly, it gave him the flexibility to be with our family, something he had longed for during his years of service.

Looking back, I'm incredibly proud of how he navigated this transition. It wasn't just a career change; it was a journey of rediscovery. He learned that stepping away from the military didn't mean losing his identity. Instead, it gave him the chance to rediscover who he was at his core, beyond the uniform.

Now, he shares his story with other veterans, reminding them that it's okay to take their time, to reflect, and to find a path that truly resonates with their strengths and values. As he often says, "After 26 years in the Air Force, I didn't need just a job. I needed to find a purpose that aligned with who I am now—and that's worth taking the time to discover."

Action Steps for Self-Assessment:

☐ **Reflect on Your Strengths**

Identify three moments from your military service where you felt proud or successful. Write down the strengths you used in each situation (e.g., leadership, problem-solving, resilience).

☐ **Take a Strengths Inventory**

Make a list of the key strengths you've identified from

your experiences. Highlight the ones that you feel are most relevant to your civilian life.

☐ **Recognize Your Skills**

Create a list of both technical and soft skills gained during your service. Identify the ones you excel at and consider how they can be applied in civilian roles.

☐ **Define Your Values**

Reflect on what truly matters to you—both personally and professionally. Write down the values that guide your decisions and actions (e.g., loyalty, integrity, service).

☐ **Align Strengths, Skills, and Values**

Review your strengths, skills, and values, and look for connections between them. Consider how these can shape your future career and life path.

☐ **Evaluate Your Weaknesses**

Acknowledge areas where you might need improvement. Understanding your weaknesses is just as important as knowing your strengths, as it helps you prepare for potential challenges in your transition.

☐ **Set Personal Goals**

Using the strengths, skills, and values you've identified, set clear, actionable goals for your civilian life. Make sure they are specific and measurable (e.g., develop leadership skills in a civilian role, gain certification in a technical field).

☐ **Assess Emotional Well-being**

Take time to evaluate your emotional readiness for the transition. Identify any emotional challenges you may face and plan ways to manage them (e.g., stress management, seeking support, practicing self-care).

☐ **Seek Feedback from Others**

Reach out to trusted colleagues, friends, or mentors to gain an outside perspective on your strengths, skills, and areas for improvement. Their insights can help you better understand your self-assessment.

☐ **Document and Revisit Your Self-Assessment**

Write down the results of your self-assessment and revisit them regularly. As you progress in your transition, update your strengths, skills, values, and goals to ensure they align with your evolving experiences and aspirations.

3

FINANCIAL READINESS

"Beware of little expenses; a small leak will sink a great ship."

— Benjamin Franklin

Transitioning from military to civilian life comes with a new set of financial challenges and opportunities. But with the right preparation, you can navigate this change smoothly and build a stable financial future. This chapter will guide you through understanding your financial situation, creating a budget, leveraging your military benefits, and planning for long-term stability. Along the way, I'll share stories of veterans who have successfully managed their finances and provide interactive tools to make the process easier.

1. Understanding Your Financial Situation

. . .

Personal Snapshot Exercise

Before diving into planning, it's crucial to take a clear look at your current financial state. Here's how you can get started:

To get a clear handle on your financial situation, it's essential to break things down into key categories: income, expenses, savings, investments, and debts. This approach will give you a solid snapshot of where you stand financially, helping you make smarter decisions and plan effectively for your future.

Income Sources Start by listing every source of income you have. This could include:

- Military pay
- Military pension
- Social Security benefits
- Income from a second job or freelance work
- Rental income
- Investment income, like dividends or interest
- Any benefits or assistance, such as VA benefits or disability payments

Expense Tracking Next, take a close look at your monthly expenses. Divide them into fixed and variable costs:

- **Fixed expenses:** Rent or mortgage, utilities, insurance (health, vehicle), loan payments (student loans, personal loans)
- **Variable expenses:** Groceries, dining out, entertainment (movies, vacations), transportation (fuel, public transport)

. . .

Savings and Investments Then, identify your savings and investments. Include things like:

- Emergency fund
- Retirement accounts (401(k), IRA)
- Brokerage accounts (stocks, bonds)
- Real estate investments
- Savings accounts

Debt Management Finally, list any debts or financial obligations you're dealing with:

- Credit card debt
- Student loans
- Personal loans
- Mortgages
- Auto loans

By organizing your financial details into these categories, you'll gain a clearer understanding of where you are right now. This insight will empower you to make informed choices, set priorities, and develop a solid financial plan as you transition to civilian life.

Remember, financial stability is a journey. It's completely normal to feel overwhelmed at times, but take it step by step. Don't hesitate to seek guidance from financial advisors or trusted mentors who've been through similar transitions.

2. Creating a Civilian Budget

Transitioning to civilian life often involves significant changes in both income and expenses. A well-structured budget is essential for adapting to these changes and maintaining financial control. This section provides a practical guide to creating a civilian budget that aligns with your new circumstances.

2.1 Civilian Budget Planner

Estimate Your New Income

Start by estimating your new sources of income. This includes your expected salary, VA benefits, and any additional income streams. Consider the following components:

- **Expected Salary:** Research the average salary for your desired job in the civilian workforce.
- **VA Benefits:** Include any benefits you will receive from the Department of Veterans Affairs.
- **Other Income:** Factor in any additional income sources, such as freelance work, rental income, or part-time jobs.

For example, if you anticipate an annual salary of $60,000, receive $1,500 monthly in VA benefits, and earn an additional $200 monthly from freelance work, your monthly income would be calculated as follows:

- Monthly Income Calculation:
  - Expected Salary: $60,000 / 12 = $5,000
  - VA Benefits: $1,500

- Other Income: $200

TOTAL MONTHLY INCOME = $5,000 + $1,500 + $200 = **$6,700**

ADJUST **Your Expenses**

NEXT, assess your expenses and make necessary adjustments based on your new civilian lifestyle. Key areas to consider include:

- **Housing Costs:** Determine your new rent or mortgage payments based on your location.
- **Healthcare Expenses:** Consider changes in health insurance premiums and out-of-pocket costs compared to military healthcare.
- **Commuting Costs:** Estimate transportation expenses, including fuel, public transit, or vehicle maintenance.

Create a detailed list of your expected monthly expenses, categorizing them into fixed and variable costs:

- **Fixed Expenses:** Rent/mortgage, utilities, insurance premiums, loan payments
- **Variable Expenses:** Groceries, dining out, entertainment, personal care

SET **Priorities**

. . .

WITH A CLEARER UNDERSTANDING of your income and expenses, prioritize your financial goals. Identify essential expenses that must be covered each month, such as:

- Housing (rent/mortgage)
- Utilities
- Groceries
- Insurance
- Minimum debt payments

Next, look for areas where you can cut back to ensure your budget aligns with your goals, such as:

- **Debt Repayment:** Allocate a portion of your income to pay down credit card debt or loans more aggressively.
- **Savings for a Home:** Set aside funds for a down payment or closing costs if homeownership is a goal.
- **Emergency Fund:** Aim to build an emergency fund that covers three to six months' worth of living expenses.

BY FOLLOWING this structured approach to budgeting, you will be better equipped to manage your finances during your transition to civilian life. This clarity will help you make informed decisions, reduce financial stress, and work toward achieving your long-term financial goals.

3. Maximizing Your Military Benefits

Understanding Your Benefits

Your military benefits are more than just entitlements—they are foundational elements of your financial well-being. Properly understanding and leveraging these benefits can have a lasting impact on your financial security.

. . .

3.1 Key Benefits to Explore

• **Retirement Benefits:** It's crucial to understand how your military pension is calculated and how it fits into your overall income strategy. This knowledge helps you plan for a comfortable retirement.

• **VA Benefits:** VA benefits offer a wide range of support, including healthcare, disability compensation, education benefits, and home loans. Familiarizing yourself with these options ensures you're making the most of what's available to you.

• **Health Insurance:** As you transition from active duty, comparing post-service healthcare options is vital. Understanding the differences between available plans helps you choose the one that best meets your needs.

Personalization for Veterans with Disabilities

For veterans with disabilities, it's important to recognize how disability compensation can influence your overall financial plan. Tailoring your approach to your unique circumstances can provide greater financial stability and peace of mind.

4. Planning for Long-Term Financial Stability

Long-Term Planning Exercise

Financial readiness extends beyond addressing your current needs—it's about ensuring long-term stability and security. Proper planning today can safeguard your financial future, providing peace of mind as you navigate life after military service.

- **Assess Your Financial Goals:** Begin by identifying your long-term financial goals, such as retirement, homeownership, or education for your children. Knowing what you're working towards helps guide your planning efforts.
- **Create a Comprehensive Financial Plan:** Develop a plan that covers all aspects of your finances, including savings, investments, insurance, and estate planning. A well-rounded approach ensures you're prepared for whatever the future holds.
- **Consider Future Healthcare Needs:** Healthcare costs can be a significant financial burden, especially as you age. Evaluating your long-term healthcare options, including long-term care insurance, can help you manage these potential expenses.
- **Review and Adjust Regularly:** Your financial situation and goals may evolve over time. Regularly reviewing and adjusting your plan ensures it remains aligned with your current needs and future aspirations.

5. Integrating Financial Readiness into Your Transition

Ongoing Financial Management

Financial readiness isn't a one-time task—it's an ongoing process that should evolve with your life circumstances. As you transition out of military service, maintaining and adjusting your financial plan is essential to ensure it continues to meet your needs.

- **Regular Financial Check-Ins:** Make it a habit to review your financial situation regularly. This allows you to stay on top of changes in your income, expenses, and financial goals, ensuring your plan remains effective.

• **Adapt to Life Changes:** Major life events, such as marriage, the birth of a child, or buying a home, can significantly impact your financial plan. Adjusting your strategy to accommodate these changes helps you stay on track.

• **Stay Informed About Benefits:** As policies and benefits evolve, staying informed about the latest updates ensures you're maximizing what's available to you. Regularly review your benefits and consult with a financial advisor if needed.

• **Emergency Preparedness:** Life is unpredictable, so it's crucial to have an emergency fund and a plan for unexpected expenses. This financial cushion provides security and peace of mind during challenging times.

By consistently managing your finances and adapting to changes, you'll be better equipped to navigate your transition and maintain long-term financial health.

Sarah's Financial Makeover: From Military Paychecks to Civilian Stability

Sarah didn't realize how much financial control she had lost—until she was drowning in debt. The transition from military paychecks to civilian life was rougher than she expected. One minute, she was confident in her ability to manage her finances; the next, she was struggling to figure out where her money was going. That's when she came to me—Ms. Veteran Affairs—desperate for guidance.

We sat down and took a hard look at her spending. One glaring issue immediately jumped out: her daily coffee shop habit. It seemed harmless enough, but when we did the math, it had snowballed into hundreds of dollars a month. I could see the disbelief in her eyes. "How did I let this happen?" she asked. But instead of dwelling on that, we focused on a solution. By cutting back on those daily treats, Sarah freed up cash to pay down her credit card debt—a huge relief for her already tight budget.

But a quick fix wasn't going to cut it. Sarah needed a financial overhaul. I encouraged her to download a budgeting app and start from scratch. Together, we mapped out her income—including her new job salary and VA benefits—and adjusted her budget for civilian expenses like healthcare and commuting. It wasn't easy, but with each adjustment, she felt a little more empowered.

"This feels different," she said, and I could tell she was starting to believe it. She finally had a plan.

Once Sarah had a grip on her day-to-day finances, we moved on to the bigger picture. She hadn't fully tapped into her VA benefits, and it was costing her opportunities. I set up a meeting with a VA representative, and Sarah came out of it more informed and energized than ever. She'd discovered education programs that could sharpen her skills and improve her job prospects. "I had no idea these resources were available," she told me, her eyes lighting up.

With that newfound knowledge, Sarah wasted no time. She enrolled in a certification program that not only expanded her skills but also opened doors to career paths she hadn't considered before. "This could change everything," she said, with a sense of optimism that had been missing for months.

And it did. Armed with both a clearer financial plan and career goals, Sarah began thinking long-term. Together, we talked about retirement savings, buying her first home, and creating an emergency fund—steps that would secure her family's future. She made it a priority to check in on her financial health regularly, adjusting her budget as needed and consulting experts to stay on track.

Months later, Sarah looked back on her journey with pride. "I used to feel so lost," she admitted, but now, with careful planning and a steady hand guiding her, she had gained financial stability and confidence. Her journey wasn't just about managing money; it was about building a foundation for the life she and her family deserved.

Action Steps for Financial Readiness

1. Understanding Your Financial Situation

☐ **Create a Financial Snapshot**

List all sources of income, including military pay, pension, VA benefits, and any additional income.

☐ **Track Your Expenses**

Categorize your expenses into fixed (rent, insurance) and variable (groceries, entertainment).

☐ **Identify Your Savings and Investments**

Detail your emergency fund, retirement accounts, brokerage accounts, and any other investments.

☐ **List Your Debts**

Include credit card debt, student loans, personal loans, mortgages, and auto loans.

☐ **Assess Your Financial State**

Compile the above information to understand your current financial position.

2. Creating a Civilian Budget

☐ **Estimate Your New Income**

Calculate your expected monthly income from salary, VA benefits, and other sources.

□ **Adjust Your Expenses**

Evaluate and update your expenses based on your new civilian lifestyle. Consider housing, healthcare, and commuting costs.

□ **Set Financial Priorities**

Identify essential expenses (housing, utilities, groceries) and allocate funds for debt repayment, savings, and emergency funds.

3. Maximizing Your Military Benefits

□ **Understand Your Retirement Benefits**

Review how your military pension fits into your overall income strategy.

□ **Explore VA Benefits**

Familiarize yourself with available VA benefits, including healthcare, disability compensation, education benefits, and home loans.

□ **Evaluate Health Insurance Options**

Compare post-service healthcare plans to find the one that best meets your needs.

4. Planning for Long-Term Financial Stability

□ **Identify Long-Term Financial Goals**

Determine goals such as retirement, homeownership, or education.

□ **Create a Comprehensive Financial Plan**

Include savings, investments, insurance, and estate planning in your plan.

□ **Consider Future Healthcare Needs**

Evaluate long-term healthcare options, including long-term care insurance.

□ **Review and Adjust Regularly**

Regularly update your financial plan to reflect changes in your situation and goals.

5. Integrating Financial Readiness into Your Transition

□ **Conduct Regular Financial Check-Ins**

Review your financial situation periodically to ensure your plan remains effective.

□ **Adapt to Life Changes**

Adjust your financial strategy for major life events (marriage, children, home purchases).

□ **Stay Informed About Benefits**

Keep up with updates on policies and benefits to maximize what's available to you.

□ **Prepare for Emergencies**

Maintain an emergency fund and have a plan for unexpected expenses.

By following these steps, you can navigate your financial transition with greater confidence and stability, paving the way for a secure and successful future.

4

OPPORTUNITIES EXPLORATION

"Opportunities are usually disguised as hard work, so most people don't recognize them."

— Ann Landers

Transitioning from military to civilian life presents a wealth of opportunities, but recognizing and seizing them requires effort, determination, and a strategic approach. This chapter will guide you through exploring career opportunities that align with your military experience, skills, and personal interests using the S.O.A.R. framework.

1. Identifying Career Opportunities

Your military service has endowed you with a unique skill set. The key is to identify how these skills can translate into fulfilling civilian careers.

Exercise: Mapping Military Skills to Civilian Careers

. . .

List Your Skills:

- **Roles and Responsibilities:** Document your military roles, responsibilities, and the skills you developed. Include both technical skills (e.g., IT proficiency) and soft skills (e.g., leadership, teamwork).
- **Example:** If you were a combat medic, your skills in emergency response and patient care are highly transferable to roles in healthcare.

Research Applications:

- **Skill Translation:** Research how these skills are valued in civilian industries. For instance, leadership experience in the military can be highly valuable in management roles across various sectors.
- **Example:** A logistics specialist in the military might find opportunities in supply chain management or operations in industries like healthcare or manufacturing.

Align with Interests:

- **Passion and Goals:** Reflect on your personal interests and how they align with potential careers. Are there fields you are passionate about that could benefit from your unique skills?

- **Example:** If you have a passion for technology, explore roles in IT or cybersecurity.

2. Researching Potential Careers

Once you've identified potential career paths, it's essential to conduct thorough research to make informed decisions.

Steps to Conduct Career Research

Job Market Analysis:

- **Resources:** Use resources like the Bureau of Labor Statistics, industry reports, and job boards to assess demand, salary ranges, and regional job availability.
- **Tools:** Websites like LinkedIn and Glassdoor can provide insights into job roles, company reviews, and salary expectations.

Education & Training:

- **Qualifications:** Determine if additional education or certifications are needed. Many veterans find that their military training provides a strong foundation, but some careers may require further education or specialized training.

- **Example:** For a career in project management, you might need to obtain a PMP (Project Management Professional) certification.

Growth Potential:

- **Prospects:** Evaluate the long-term prospects of your chosen careers. Are these fields growing, stable, or declining? What opportunities exist for career advancement?
- **Example:** Consider industries like technology and healthcare, which are known for strong growth and advancement opportunities.

3. Building Your Professional Network

A strong professional network is crucial for your career exploration and job search.

Strategies for Building Your Network

Leverage Military Connections:

- **Fellow Veterans:** Reach out to fellow veterans who have successfully transitioned into civilian careers. They can offer valuable advice, referrals, and support.

- **Example:** Join veteran-specific groups or online forums where you can connect with others who have made similar transitions.

Attend Industry Events:

- **Networking Opportunities:** Participate in job fairs, industry conferences, and local networking events. These settings allow you to meet potential employers, learn about industry trends, and make valuable connections.
- **Tip:** Prepare a brief personal pitch to introduce yourself effectively.

Join Professional Associations:

- **Industry Groups:** Consider joining associations related to your field. These organizations often provide resources like job boards, training programs, and networking events.
- **Example:** If you're interested in finance, join the CFA Institute or similar organizations.

Utilize Social Media:

- **LinkedIn:** Engage on platforms like LinkedIn. Join relevant groups, follow industry leaders, and share your experiences to build a strong online presence.

- **Tip:** Regularly update your LinkedIn profile and engage with content to stay visible in your field.

Your Military Experience: A Collection of Past Achievements or the Key to Your Next Career?

When Sergeant Mike Thompson, a close friend of my late husband, faced the daunting task of transitioning from 15 years of service in the Army to civilian life, he wasn't just dealing with a change in career—he was grappling with a profound shift in identity. Mike was known for his leadership, his ability to manage logistics with precision, and his knack for solving complex problems under pressure. But as he stared at a stack of discharge paperwork one evening, he felt lost, unsure how to translate his military achievements into a meaningful civilian career.

I remember the night he called me, overwhelmed and uncertain. "How did Eddie do it?" he asked, his voice tinged with frustration and worry. Eddie, my husband, had faced similar challenges, and I had walked with him through the ups and downs of that transition. "Mike," I told him, "Your military experience is more than just a collection of past achievements. It's the foundation for your next great career. You just need to unlock its potential."

We started by taking a deep dive into his skills. I helped him see that his leadership, strategic thinking, and problem-solving abilities weren't just military assets—they were powerful tools that any industry would value. Together, we explored how these skills could be applied in civilian sectors. Through our conversations, Mike identified fields like healthcare, technology, and manufacturing, where his background could truly shine.

But it wasn't just about identifying opportunities. I knew that for Mike to succeed, he needed to expand his network. I encouraged him to attend conferences, join professional associations, and connect with other veterans who had made similar transitions. These connections not only provided insights but also opened doors to new job opportunities.

Realizing there were gaps in his knowledge, I guided Mike on how to use his GI Bill benefits to enroll in online courses for IT project management—an area where his skills were particularly relevant. It wasn't easy for him to balance studies with family responsibilities, but with determination, Mike powered through, completing the certifications that would set him apart in his new career path.

With a clear plan and unwavering support, Mike set out to achieve his goals: securing a project management role within six months, completing his certification, and building a strong professional network. He broke these goals down into actionable steps, sought mentorship from industry experts, and remained committed, even when the road got tough.

Mike's efforts paid off. Within months, he landed a role as an IT Project Manager, where his military experience and leadership skills were not just acknowledged—they were celebrated. His journey is a testament to the power of strategic planning, the importance of a solid support system, and the incredible potential that lies within every veteran's experience.

So, if you find yourself standing at a similar crossroads, remember Mike's story. With the right strategy, support, and determination, your military experience can be the key to unlocking a fulfilling new career. And just like Mike, you don't have to walk that path alone.

Action Steps Checklist for Chapter 5: Opportunities Exploration

- **Identify Your Skills**
- List your military roles, responsibilities, and the technical and soft skills you developed.
- **Research Civilian Applications**
- Research how your military skills translate into civilian career opportunities.
- **Align Skills with Interests**
- Reflect on your personal interests and align them with potential career paths.
- **Conduct Job Market Analysis**
- Use resources like the Bureau of Labor Statistics and job boards to assess demand, salary, and job availability.
- **Explore Education & Training Requirements**
- Determine if your chosen career path requires additional education or certifications.
- **Evaluate Career Growth Potential**
- Research the long-term prospects and growth opportunities in your chosen fields.
- **Leverage Your Military Network**
- Reach out to fellow veterans for advice and support in your transition.

- **Attend Networking Events**
- Participate in job fairs, industry conferences, or local networking events.
- **Join Professional Associations**
- Identify and join industry groups related to your career interests.
- **Build Your LinkedIn Presence**
- Update your LinkedIn profile and engage with industry content regularly.
- **Develop a Networking Plan**
- List key contacts in your desired industry and create an outreach plan.
- **Utilize Your GI Bill Benefits**
- Research educational programs and institutions covered by your GI Bill benefits.
- **Enroll in Necessary Courses or Certifications**
- Identify and start any required education or certifications for your new career.
- **Set Clear Career Goals**
- Define success in your career transition, such as securing a job, completing certifications, or expanding your network.
- **Break Down Action Items**
- Create specific action steps to achieve your career goals.
- **Establish Timelines**
- Set realistic deadlines for completing each action step.
- **Seek Mentorship**
- Find career mentors or counselors to support and guide your transition.

These steps will help you strategically explore and seize new career opportunities. Check them off as you progress through your career transition journey.

5

UNDERSTANDING YOUR BENEFITS

"The best way to predict your future is to create it."

— Abraham Lincoln

As you transition from military to civilian life, understanding and leveraging your benefits is crucial to shaping your future. This chapter explores how to make the most of your military benefits, integrating the Opportunities Exploration aspect of the S.O.A.R. framework to support your career and educational goals. By grasping what's available, you can make informed decisions that pave the way for a successful transition.

1. Navigating Your Benefits

Veterans are entitled to a range of benefits designed to support their transition and future endeavors. Here's how to navigate and utilize them effectively:

. . .

Types of Benefits Available

Education Benefits:

- **GI Bill:** The Post-9/11 GI Bill and Montgomery GI Bill provide financial support for education and training, covering tuition, fees, and housing.
- **Vocational Rehabilitation and Employment (VR&E):** Chapter 31 helps veterans with service-connected disabilities achieve their employment goals through education and training

Healthcare Benefits:

- **VA Health Care:** Access a comprehensive range of medical services, including primary care, mental health services, and specialist care.

Home Loans:

- **VA Home Loan Program:** Assists veterans in purchasing, building, or refinancing a home with favorable terms, including no down payment and competitive interest rates.

Other Benefits:

- **Education and Career Counseling:** Offers guidance on educational and career choices, helping you plan your next steps.
- **Disabled Veterans' Benefits:** Provides additional support for veterans with service-connected disabilities, including financial compensation and support services.

How to Access Your Benefits

Research Available Programs:

- **Example:** Sarah, a veteran transitioning to civilian life, used the VA's official website to explore the Post-9/11 GI Bill. She discovered it could cover her tuition for a degree in computer science and provide a monthly housing allowance.

Apply for Benefits:

- Complete the necessary application forms online or through a VA representative. For example, Sarah applied for the GI Bill through the VA portal and received approval within weeks.

Consult with a Benefits Advisor:

- Schedule an appointment with a VA benefits advisor for personalized assistance. They can help you navigate the application process and maximize your benefits.

2. Integrating Benefits into Your Career and Educational Goals

With a clear understanding of your benefits, the next step is integrating them into your career and educational plans.

Aligning Benefits with Career Goals

Education and Training:

- **Example:** John, an Army veteran interested in project management, used his GI Bill benefits to enroll in a certification course. This helped him transition into a project management role at a tech company.

Networking and Mentorship:

- Leverage VA-sponsored networking events and mentorship programs. John attended industry conferences to connect with professionals and gain insights into his new field.

Planning for Financial Stability

- **Budgeting:**
- Create a budget that includes your benefits and other income sources. For instance, John budgeted his monthly housing allowance from the GI Bill alongside his salary to manage his finances effectively.
- **Emergency Funds:**
- Set aside funds for unexpected expenses. John established an emergency fund to cover any unforeseen costs during his transition.

Utilizing Healthcare Benefits

- **Accessing Medical Services:**
- Regularly use VA healthcare services to maintain your well-being. John accessed mental health services provided by the VA to support his transition.
- **Understanding Coverage:**
- Familiarize yourself with the coverage under your VA health benefits. Knowing what's included helps you make the most of the services available.

3. Making Informed Decisions

Understanding your benefits is crucial for making informed decisions about your future. Here's how to approach this process:

Research and Plan

- **Evaluate Your Needs:**
- **Example:** Lisa, a veteran planning to start a small business, used her education benefits to take courses in entrepreneurship. She evaluated her needs and aligned her benefits to support her new venture.
- **Explore All Options:**
- Don't limit yourself to one type of benefit. Combine education benefits with networking opportunities to enhance your career prospects. Lisa attended VA-sponsored networking events to connect with other entrepreneurs.

Seek Guidance

- **Consult with Experts:**
- Reach out to veterans' service organizations, educational counselors, and career advisors for personalized advice. They can provide valuable insights and help you make the most of your benefits.
- **Stay Informed:**

- Benefits and programs can change over time. Stay updated on any changes by regularly visiting the VA's official website and subscribing to veterans' newsletters.

Finding Your Path: A Veteran's Journey to Accessing the Benefits You Deserve

In my work with veterans, I've seen firsthand how understanding your benefits can transform lives. Take, for example, a remarkable young woman I had the honor of helping last year. After her honorable discharge from the military, she found herself grappling with hearing loss, vertigo, depression, and sleep apnea. Having worked on the flight line, she wasn't sure if her relatively short service would qualify her for any benefits.

When she first came to me, she was uncertain and hesitant. She had been dealing with these conditions in silence, never thinking to apply for help. But I knew that her situation was likely service-connected, and I assured her that she wasn't alone in this. "Your service matters, and so do you," I told her. My goal was to help her understand the full extent of the benefits she had earned through her service.

Together, we went through the details of her medical history and service record. I explained that her hearing loss, tinnitus, and other health issues were not just random occurrences—they were conditions that the VA recognized as service-connected disabilities. I also highlighted how the Aid and Attendance program could support her spouse, who was providing daily care for her.

With this new understanding, she felt empowered to take the next steps. We carefully navigated the application process, ensuring that all her conditions were thoroughly documented. To her surprise and relief, not only did she qualify for disability compensation, but her spouse was also eligible for benefits through the Aid and Attendance program.

Her story is a powerful reminder of why understanding your benefits is so crucial. Many veterans, like her, don't realize the full range of support available to them. But with the right guidance and knowledge, those benefits can make a world of difference.

If you're feeling uncertain about what you might be entitled to, take a page from her story. Don't hesitate to explore your options and seek out the support you need. Your service has earned you these benefits, and they're here to help you build a better future.

Action Steps for Understanding Your Benefits

1. Navigating Your Benefits

☐ **Research Available Programs**

Explore benefits like the GI Bill, VA Health Care, and VA Home Loan Program through official VA resources.

☐ **Apply for Benefits**

Complete application forms online or with a VA representative for benefits like education, healthcare, or home loans.

☐ **Consult with a Benefits Advisor**

Schedule an appointment with a VA benefits advisor for personalized guidance and assistance.

2. Integrating Benefits into Your Career and Educational Goals

□ **Align Benefits with Career Goals**

Utilize education benefits for training and certification relevant to your career goals. Explore networking and mentorship opportunities.

□ **Plan for Financial Stability**

Create a budget including your benefits and other income. Set up an emergency fund for unexpected expenses.

□ **Utilize Healthcare Benefits**

Regularly access VA healthcare services and understand your coverage to maintain your well-being.

3. Making Informed Decisions

□ **Research and Plan**

Evaluate your needs and explore all available benefits. Align them with your career and educational plans.

□ **Seek Guidance**

Consult with veterans' service organizations, educational counselors, and career advisors for tailored advice.

□ **Stay Informed**

Keep updated on benefit changes by visiting the VA's official website and subscribing to newsletters.

By following these steps, you'll effectively navigate and leverage your benefits to support a successful transition to civilian life.

6

ACCESSING SUPPORT SERVICES

"Alone we can do so little; together we can do so much."

— Helen Keller

Transitioning from military life to civilian life isn't a solo mission—it's a team effort. Sure, you've laid out your game plan with the Opportunities Exploration and Action Plan, but to truly succeed, you'll need to tap into the support systems around you. This chapter is all about how to connect with the right people, resources, and services that'll help you stay on track, reach your goals, and keep your well-being front and center.

1. Why Support Services Are Game-Changers

Let's face it: civilian life is a whole new world with its own set of challenges. But the good news is, you don't have to go it alone. Support services are designed to help you every step of the way—whether it's finding a new job, managing your mental health, or navigating the benefits system. The trick is knowing what you need and where to find it.

2. Building Your Dream Team

Having a strong support network is like having a safety net during your transition. Your team can be made up of fellow veterans, family, mentors, and community groups. These are the people and organizations that understand where you're coming from and can offer the advice and encouragement you need.

How to Build Your Support Network:

- **Veteran Networking Events:**
- Whether it's local meetups or virtual hangouts, attending these events gives you a chance to connect with people who've been through what you're going through. You'll find valuable tips, job leads, and maybe even a mentor.
- **Online Veteran Communities:**
- Platforms like LinkedIn and specialized veteran forums are great for sharing stories, advice, and resources. Think of it as a virtual squad that's there to back you up whenever you need it.
- **Local Veteran Organizations:**
- Don't overlook organizations like the American Legion or VFW. They offer a wealth of resources —everything from job search help to mental health support.

3. Career Counseling: Making the Most of Your Military Skills

Your military service has equipped you with a wealth of skills, but transitioning them into civilian jobs can be challenging. **Career counseling services** are here to bridge that gap and ensure you find a career that's both fulfilling and aligned with your strengths. Here's how they can help:

Assess Your Skills

Career counselors will work with you to identify your transferable skills—those abilities and experiences from the military that are valuable in civilian roles. They can help you see how your skills in leadership, project management, and problem-solving apply to various industries.

Explore Career Paths

Not sure where your skills fit? Career counselors can introduce you to different career options that match your interests and experience. They can provide insights into industries you might not have considered and help you understand what roles could be a great fit for you.

Develop a Tailored Plan

A career counselor will help create a personalized career plan. This includes:

- **Resume Building:** Crafting a resume that highlights your military experience in a way that resonates with civilian employers.
- **Interview Coaching:** Preparing you for interviews by focusing on how to articulate your military experience and skills effectively.
- **Job Search Strategies:** Offering tips on where and how to find job opportunities, including networking and job boards.

Access Specialized Resources

Many career counseling services offer additional resources such as:

- **Workshops and Seminars:** Attend workshops on job search strategies, resume writing, and interview techniques.
- **Networking Events:** Connect with industry professionals and other veterans through career fairs and networking events.
- **Online Tools:** Utilize tools and platforms that can help with job searches, skill assessments, and career planning.

Seek Guidance and Feedback

Regular meetings with a career counselor can provide ongoing support and feedback. They can help you stay motivated and adjust your career plan as needed, ensuring that you're on track to meet your goals.

Connect with Local and Online Resources

- **Veteran Employment Programs:** Check out veteran-specific employment programs and resources that can offer additional support.
- **Professional Associations:** Join associations related to your field of interest to gain industry insights and expand your professional network.

4. Prioritizing Your Mental Health

The emotional side of transitioning can't be ignored. You might feel lost, anxious, or even frustrated, and that's completely normal. What matters most is recognizing when you need a hand and knowing that there's no shame in asking for help. In fact, reaching out for support shows strength and self-awareness.

The VA offers a variety of mental health services specifically designed for veterans, from counseling to therapy and psychiatric care. These services are tailored to help you navigate the unique challenges that come with leaving the military. If you're feeling overwhelmed or simply need someone to talk to, the VA can connect you with professionals who understand what you're going through.

Local *Vet Centers* are another great option. They provide free counseling to veterans and their families, with a focus on military-related issues like PTSD, anxiety, and the stress of reintegration into civilian life. These centers often have a more personal, community feel, which can be a relief if you're looking for a less formal setting.

Nonprofit organizations also have your back. Groups like *Give an Hour* or the *Wounded Warrior Project* specialize in veteran mental health and can connect you with licensed therapists at no cost. Many of these services are available in person, but there's also a growing number of telehealth options if you prefer talking to someone from the comfort of home.

For veterans who want flexibility, teletherapy might be a perfect fit. Services like *VA TeleMental Health* or platforms like *Talkspace* offer virtual sessions, making it easier to fit mental health care into your schedule. It's a simple, effective way to get the support you need, no matter where you are.

And if you're more comfortable talking with fellow veterans, consider joining a support group—whether in-person or online. Sometimes, just sharing your story with people who've walked a similar path can lift some of the weight off your shoulders.

Remember, the small things count too. Simple self-care practices like meditation, mindfulness, or even exercise can make a big difference in how you feel day-to-day. Apps like *Headspace* and *Calm* offer guided meditations specifically for managing stress and anxiety.

If things ever get too heavy, the *Veterans Crisis Line* is available 24/7. It's okay to reach out when things feel like too much—there's always someone ready to listen.

5. Community Resources: Local Help When You Need It

Sometimes, the best help is right in your backyard. Local communities often have a treasure trove of resources—everything from housing assistance and job training to healthcare services. These programs are built with you in mind, offering support that's tailored to your unique situation. Whether you need immediate help or long-term guidance, there's likely a service nearby that can step in.

Start by checking out community centers, local nonprofits, and even city or state government offices. Many of these organizations are veteran-friendly and ready to help you transition smoothly into civilian life. Sometimes, just asking the right questions will lead you to a service you didn't even know existed.

From Uncertainty to Clarity: How Support Services Made a Difference

I'd like to share a story about Tom Reed, an Air Force veteran whose journey through transition is a testament to the power of accessing the right support services. This story is not just about Tom, but also resonates deeply with my own experiences and those of my family.

Tom served as a communications specialist for 15 years, and like many veterans, he faced significant challenges when transitioning to civilian life. His skills were impressive, but translating them into a civilian career felt overwhelming. Recognizing the need for support, Tom reached out to our veterans' resource center. As someone who has also navigated these challenges, I understood his struggles and was committed to helping him find clarity.

Tom's first step was career counseling. I connected him with a career counselor who specializes in working with veterans. They worked together to translate Tom's military experience into a civilian context. For Tom, this meant seeing how his expertise in managing communication systems could apply to IT project management. The counselor's guidance was instrumental in refining Tom's resume and preparing him for interviews, which was a turning point for him.

Understanding the emotional toll that transitions can take, I recommended that Tom access our mental health services. Like many veterans, Tom felt stressed and overwhelmed. I encouraged him to attend weekly counseling sessions and join a support group for veterans. Having faced similar challenges with my own family, I knew how vital these resources could be. The support group became a crucial part of Tom's journey, offering a space to share experiences and strategies for managing stress.

I also introduced Tom to various community resources. Through our center, Tom learned about a job training program that offered IT certifications. Thanks to a veterans' scholarship, he was able to enroll and gain the qualifications he needed. Additionally, he accessed a community housing assistance program that provided transitional housing while he searched for a permanent home. These resources helped Tom find stability and direction, much like the support my family and I benefited from during our own transition.

Networking was another important aspect of Tom's journey. I encouraged him to attend veterans' networking events organized by our center. At one such event, Tom met Lisa, a veteran who had successfully transitioned into IT. Lisa's advice and connections were invaluable, providing Tom with job leads and insights. Seeing Tom's progress reminded me of how crucial these networking opportunities were for my own family as well.

Finally, I showed Tom how to use online tools like the VA's Vet Centers Locator. These tools are designed to help veterans find additional resources and stay informed about new opportunities. They were essential for Tom as he navigated his transition, just as they were for us.

Tom's story is a powerful reminder of the impact that accessing the right support services can have. By leveraging career counseling, mental health support, community resources, networking opportunities, and interactive tools, Tom was able to navigate his transition successfully. His journey mirrors many of the experiences I've seen, including my own, and underscores the importance of reaching out for help and utilizing available resources.

If you're navigating your own transition, remember that support is available and can make a significant difference. Just like Tom, you too can find clarity and direction with the right guidance and resources.

HERE'S a quick checklist to help you explore local resources:

☑ **Community Centers:**

- Visit your local community center and ask about veteran-specific programs. These centers often offer workshops, job training, or social services that are catered to veterans.

☑ **Nonprofit Organizations:**

- Reach out to local nonprofits. Look for those that focus on housing assistance, career development, or healthcare for veterans. Some may offer free legal aid or help with disability claims.

☑ **Government Services:**

- Don't forget about city, county, or state government resources. Many provide housing vouchers, employment services, and healthcare benefits tailored to veterans. Look for veteran service officers who can guide you through accessing these benefits.

☑ **Veteran Service Organizations (VSOs):**

- Check with organizations like the VFW, American Legion, or local VSOs for additional resources. They often have connections to community support programs and can advocate on your behalf.

☑ **Local Job Training Programs:**

- Investigate nearby colleges, vocational schools, or community organizations offering job training programs. Many provide scholarships or veteran-preferred admission.

☑ **Healthcare Clinics & Mental Health Support:**

- Look for local health clinics, mental health centers, or telehealth options. Many communities offer veteran-specific healthcare services outside the VA.

☑ **Networking & Social Groups:**

- Join local veteran social groups or networking clubs. These informal settings often lead to hearing about new resources and job opportunities.

7

DISABILITY AND COMPENSATION

"A hero is someone who has given his or her life to something bigger than oneself."

— Joseph Campbell

Navigating disability benefits and compensation can be one of the most critical aspects of your transition from military to civilian life. Understanding and securing the benefits you're entitled to can significantly impact your ongoing care and overall well-being. This chapter will provide you with a comprehensive guide to the process, offering practical advice, interactive tools, and emotional support to help you succeed.

Understanding Disability Benefits and Compensation

. . .

Disability benefits are designed to assist veterans who have experienced injuries or illnesses due to their service. These benefits typically include:

- **Financial Assistance:** Monthly payments to help with living expenses.
- **Healthcare Access:** Coverage for medical treatments and prescriptions.
- **Additional Support:** Programs for rehabilitation, vocational training, and more.

Why It Matters

Securing these benefits is not just about financial support—it's about ensuring you have the resources to manage your health, maintain your quality of life, and focus on your personal and professional goals.

Step-by-Step Process

Gather Documentation

Start by compiling all necessary documents:

- **Medical Records:** Detailed accounts of your injuries or conditions.
- **Service Documents:** Proof of your military service and its impact on your health.
- **Supporting Evidence:** Statements from medical professionals and fellow service members.

File Your Claim

- **Veterans Service Officer (VSO):** Seek help from a VSO who can guide you through the VA's eBenefits portal and ensure your claim is complete and accurate.
- **Submission:** Submit your claim with the necessary documentation, ensuring all information is correct.

Attend VA Appointments

- **Medical Evaluations:** Participate in required evaluations and hearings to verify your disability and eligibility for benefits.
- **Follow-ups:** Keep track of all appointments and ensure you follow through on any recommended treatments or evaluations.

Appeal if Necessary

- **Initial Denial:** If your claim is denied or partially approved, gather additional evidence and seek guidance from your VSO.
- **Appeal Process:** File an appeal with the new evidence. Persistence and proper documentation are key.

Managing Ongoing Care and Support

Once approved for benefits, effective management of your ongoing care is crucial. This includes:

- **Regular Medical Appointments:** Stay up-to-date with your healthcare provider for ongoing treatments.
- **Physical Therapy:** Follow through with prescribed physical therapy to aid in recovery.
- **Mental Health Support:** Access counseling or therapy if needed, to address any emotional or psychological challenges.

Utilizing Resources for Additional Support

Beyond VA benefits, various organizations offer specialized support for veterans with disabilities:

- **Nonprofit Organizations:** Groups like *Wounded Warrior Project* and *Disabled American Veterans* offer additional assistance and resources.
- **Community Health Programs:** Explore local programs that provide healthcare, rehabilitation, and vocational support.
- **Specialized Veteran Services:** Look into services tailored to your specific needs, such as adaptive technologies or housing modifications.

Interactive Resources and Tools

Leverage online tools to manage and track your benefits effectively:

- **VA Disability Compensation Calculator:** Use this tool to estimate your benefits and plan your finances.
- **eBenefits Portal:** Regularly check the status of your claim and manage your benefits online.
- **Veteran Support Forums:** Engage with online communities to share experiences and get advice.

Navigating the disability and compensation process is a vital part of your transition to civilian life. By understanding the benefits available to you, managing your care effectively, and utilizing additional support resources, you can build a solid foundation for a fulfilling post-military life. Remember, seeking help is a strength, not a weakness. There are numerous resources and support systems ready to assist you every step of the way.

Don't hesitate to reach out and use the available tools and support networks to ensure you receive the assistance you deserve. Your journey towards a successful and rewarding civilian life starts with taking these crucial steps.

Samuel's Path to Recovery: Securing the Benefits You Deserve

Let me share with you the journey of Samuel, a former Air Force technician whose path to securing disability benefits was a testament to resilience and the importance of accessing the right support. As someone who has worked closely with veterans, including my own family, I understand the hurdles that come with transitioning to civilian life.

Samuel came to me struggling with persistent shoulder pain from years of handling heavy equipment. His physical discomfort was affecting not only his daily activities but also his job prospects. Seeing how overwhelmed he felt, I guided him through the process of securing the benefits he deserved.

The first piece of advice I gave Samuel was to seek help from a Veterans Service Officer (VSO). These professionals are well-versed in navigating the complexities of disability claims and can provide invaluable assistance. Samuel worked closely with his VSO, who helped him gather necessary medical records, compile service documentation, and file his initial claim.

Samuel's initial claim was denied, which can be disheartening. However, I emphasized the importance of persistence. With the VSO's guidance, Samuel decided to appeal the decision. They reviewed his case, provided additional evidence, and resubmitted the claim. This persistence paid off, and Samuel eventually received the disability compensation he needed.

Once Samuel secured his benefits, he gained access to specialized treatments and a support group for veterans with similar injuries. I encourage veterans to make full use of these resources. The support group was particularly helpful for Samuel, offering him practical advice, camaraderie, and a sense of community.

In addition to disability compensation, I advised Samuel to explore local community support programs. These programs often offer job training, housing assistance, and other resources that can make the transition smoother. Samuel benefited from job training programs that provided him with additional certifications and access to transitional housing.

Finally, I encouraged Samuel to stay informed about available resources and support services. Regularly checking in with the VSO and using online tools to find additional services can help veterans stay updated on new opportunities and support.

Samuel's story underscores the importance of perseverance and utilizing available resources when navigating disability benefits. By seeking help early, persisting through setbacks, and fully exploring support programs, veterans can successfully manage their conditions and transition to civilian life.

If you're facing similar challenges, remember that support is available, and with the right guidance and determination, you can secure the benefits and resources you need. Just like Samuel, you can find a path to recovery and a smoother transition to civilian life.

Action Steps for Disability and Compensation

1. Gather Documentation

☐ **Compile Medical Records**

Collect detailed records of your injuries or conditions.

☐ **Obtain Service Documents**

Gather proof of your military service and its impact on your health.

☐ **Collect Supporting Evidence**

Secure statements from medical professionals and fellow service members.

2. File Your Claim

☐ **Seek Help from a Veterans Service Officer (VSO)**

Find a VSO to assist with navigating the VA's eBenefits portal and ensure your claim is complete.

☐ **Submit Your Claim**

File your claim with all necessary documentation and correct information.

3. Attend VA Appointments

☐ **Participate in Medical Evaluations**

Attend required evaluations and hearings to verify your disability and eligibility.

☐ **Follow-Up on Treatments**

Keep track of appointments and follow through on any recommended treatments.

4. Appeal if Necessary

☐ **Address Initial Denial**

If denied or partially approved, gather additional evidence and seek guidance from your VSO.

☐ **File an Appeal**

Submit an appeal with new evidence and ensure persistence throughout the process.

5. Manage Ongoing Care and Support

☐ **Schedule Regular Medical Appointments**

Stay up-to-date with your healthcare provider for ongoing treatments.

☐ **Follow Physical Therapy**

Complete prescribed physical therapy to aid in your recovery.

☐ **Access Mental Health Support**

Seek counseling or therapy if needed to address emotional or psychological challenges.

6. Utilize Resources for Additional Support

☐ **Explore Nonprofit Organizations**

Connect with groups like Wounded Warrior Project and Disabled American Veterans for extra assistance.

☐ **Engage with Community Health Programs**

Find local programs offering healthcare, rehabilitation, and vocational support.

☐ **Investigate Specialized Veteran Services**

Look into services tailored to your needs, such as adaptive technologies or housing modifications.

7. Use Interactive Resources and Tools

☐ **Utilize VA Disability Compensation Calculator**

Estimate your benefits and plan your finances using the online tool.

☐ **Check eBenefits Portal Regularly**

Monitor the status of your claim and manage your benefits online.

☐ **Engage in Veteran Support Forums**

Join online communities to share experiences and receive advice.

By following these steps, you'll effectively navigate the disability benefits and compensation process, ensuring you receive the support and resources you need.

8

CRAFTING YOUR ACTION PLAN

"The only limit to our realization of tomorrow
is our doubts of today."

— Franklin D. Roosevelt

Transitioning from military to civilian life is a significant change, but with a well-crafted Action Plan, you can navigate this journey smoothly. This chapter provides a step-by-step guide to help you build a successful career, manage your finances, and utilize available resources effectively.

1. Building a Resume That Works in the Civilian World

Translate Your Military Experience: Your military experience is rich with skills and achievements, but translating them into civilian terms is key.

- **Identify Transferable Skills:** Break down your military roles and responsibilities to highlight skills that are valuable in the civilian job market. For instance, leadership in the military can translate to team management or project coordination in a civilian role.
- **Use Civilian-Friendly Language:** Avoid military jargon. Instead, use industry-specific terms that potential employers will understand. For example, replace "platoon sergeant" with "operations manager" or "team leader."
- **Highlight Your Achievements:** Focus on measurable outcomes that showcase your impact. Instead of listing duties, highlight achievements like, "Led a team of 15 to complete a project three weeks ahead of schedule."

2. Conducting a Targeted Job Search

Focus Your Efforts: A well-targeted job search can save time and increase your chances of landing the right job.

- **Research Potential Employers:** Look for companies that actively hire veterans or align with your skills and interests. Industries like technology, logistics, and management often value the discipline and expertise veterans bring.
- **Leverage Your Network:** Your network is one of your most powerful tools. Engage with veteran networks, professional organizations, and LinkedIn connections. Attend industry-specific events and job fairs to expand your reach.

- **Prepare for Interviews:** Practice answering common interview questions with a focus on how your military experience can benefit civilian employers. Be ready to explain how your skills and achievements can solve problems for the companies you're targeting.

3. Building a Financial Safety Net

Plan for Financial Stability: Financial planning is essential to a smooth transition. A well-thought-out strategy can provide peace of mind and stability.

- **Create a Budget:** Develop a detailed budget that accounts for your income, expenses, and savings goals. Budgeting tools like Mint or You Need a Budget (YNAB) can help you track and manage your finances.
- **Manage Debt:** If you have existing debt, create a plan to pay it off. Consulting with a financial advisor can offer personalized strategies and help you make informed decisions.
- **Build an Emergency Fund:** Aim to save three to six months' worth of living expenses. This emergency fund will act as a buffer during your job search or any unexpected financial challenges.

4. Leveraging Veteran Services

Tap Into Available Resources: Veteran services and

support networks can make a huge difference in your transition.

- **Utilize Veteran Services:** Organizations like The American Legion, DAV, and others offer career counseling, job placement services, and more. Don't hesitate to take advantage of these resources.
- **Career Counseling:** Working with a career counselor can help you refine your job search strategy and explore new career paths that match your skills and interests.
- **Join Support Groups:** Connecting with other veterans through support groups like Veterans of Foreign Wars (VFW) or Team Rubicon can provide camaraderie and additional resources to help you succeed.

5. Overcoming Challenges

Strategize for Success: Transitioning comes with challenges, but with the right strategies, you can overcome them.

- **Adapting to a New Work Environment:** Civilian workplaces can be different from the military. Seek out mentorship or coaching to help you adjust to new norms and expectations.
- **Balancing Finances While Job Searching:** If financial stability is a concern, consider taking on temporary or part-time work while you search for a full-time position. This can alleviate

financial pressure and give you more time to find the right job.

- **Managing Stress and Mental Health:** Transitioning can be stressful. Utilize mental health resources and support groups to maintain your well-being during this period.

Crafting a comprehensive Action Plan is essential to a successful transition from military to civilian life. By focusing on building a strong resume, conducting a targeted job search, managing your finances, and leveraging available support, you can navigate this journey with confidence. Remember, each step forward brings you closer to a fulfilling civilian career. Stay proactive, use your resources, and trust in your ability to succeed. Your successful transition is within reach.

TURNING Military Skills into Civilian Success: A Veteran's Action Plan

When Dan left the Navy after 20 years, he felt both excited and anxious. He was proud of his service but unsure how to navigate the civilian world. The first thing he did was sit down and create an action plan, just like he'd done countless times during his military career.

Dan started with his resume. He had led missions, managed teams, and tackled complex logistics, but he quickly realized that civilian employers didn't speak "military." So, he took a step back, identified his transferable skills—leadership, project management, and problem-solving—and rewrote his resume in terms civilians would understand. He replaced "mission commander" with "project leader" and highlighted specific achievements, like how he improved team efficiency by 15%.

Next, Dan focused on his job search. He didn't just send out resumes randomly; he researched companies that valued veterans, used job boards tailored to vets, and attended networking events. Through a veteran connection, he learned about a job opening in a logistics company. He tailored his resume for that role and prepared thoroughly for the interview, ensuring he could explain how his military experience made him the best candidate. He got the job.

Financial planning was also a crucial part of Dan's transition. He knew his steady military paycheck was ending, so he created a detailed budget and started building an emergency fund. He consulted with a financial advisor who helped him manage his savings and plan for the future. This financial cushion gave him peace of mind during his job search.

Dan didn't do it alone. He reached out to veteran services for career counseling and joined a local support group. Talking with other veterans who had successfully transitioned helped him feel less alone and more confident in his journey.

Transitioning wasn't easy, but by creating a solid action plan, Dan moved from uncertainty to a successful new career and stable civilian life. He approached it step by step, using the same discipline and focus that served him well in the Navy. And just like in the service, he found that with the right plan and support, no mission is impossible.

Action Plan for Crafting Your Transition Strategy

1. Building a Resume That Works in the Civilian World

Goal: Create a compelling resume that effectively translates military experience into civilian job market language.

STEPS:

Review Your Military Experience:

- List all your military roles, responsibilities, and achievements.

Identify Transferable Skills:

- Break down your military roles to highlight skills such as leadership, project management, and technical expertise.

Translate Military Terms:

- Replace military jargon with civilian-friendly language. For example, change "platoon sergeant" to "operations manager."

Highlight Achievements:

- Focus on measurable outcomes and specific achievements. For instance, "Led a team of 15 to complete a project three weeks ahead of schedule."

Craft Your Resume:

- Use a professional resume template and ensure it is tailored to each job application.

Resources: Resume-building workshops, online resume builders, and professional resume writers.

2. Conducting a Targeted Job Search

Goal: Execute an efficient job search to find the right position and company fit.

STEPS:

Research Potential Employers:

- Identify companies that value veteran skills or align with your career interests.

Leverage Your Network:

- Connect with veteran networks, professional organizations, and LinkedIn connections. Attend industry-specific events and job fairs.

Prepare for Interviews:

- Practice common interview questions, emphasizing how your military experience benefits potential employers.

Apply Strategically:

- Customize your resume for each application, focusing on how your skills and experience meet the job requirements.

Resources: Job search websites, veteran job boards, networking events, and career counselors.

3. Building a Financial Safety Net

Goal: Establish financial stability to support your transition and reduce stress.

STEPS:

Create a Budget:

- Develop a detailed budget that includes income, expenses, and savings goals. Use budgeting tools like Mint or YNAB.

Manage Debt:

- Develop a plan to address and reduce any existing debt. Consult a financial advisor if necessary.

Build an Emergency Fund:

- Aim to save three to six months' worth of living expenses to act as a buffer during your transition.

Monitor and Adjust:

- Regularly review and adjust your budget and savings plan as needed.

Resources: Budgeting apps, financial advisors, and debt management programs.

4. Leveraging Veteran Services

Goal: Utilize available veteran services and support networks to enhance your transition.

STEPS:

Utilize Veteran Services:

- Reach out to organizations like The American Legion, DAV, and others for career counseling and job placement services.

Seek Career Counseling:

- Work with a career counselor to refine your job search strategy and explore new career paths.

Join Support Groups:

- Connect with other veterans through support groups such as Veterans of Foreign Wars (VFW) or Team Rubicon.

Engage with Resources:

- Explore additional resources like workshops, seminars, and online tools for veterans.

Resources: Veteran service organizations, career counselors, support groups, and veteran resource centers.

5. Overcoming Challenges

Goal: Address and manage challenges to ensure a smooth transition.

STEPS:

Adapt to a New Work Environment:

- Seek mentorship or coaching to help you adjust to civilian workplace norms and expectations.

Balance Finances While Job Searching:

- Consider temporary or part-time work to alleviate financial pressure while you search for a full-time position.

Manage Stress and Mental Health:

- Utilize mental health resources and support groups to maintain your well-being during the transition period.

Develop Resilience Strategies:

- Implement stress management techniques and stay proactive in seeking support when needed.

Resources: Mental health services, temporary employment agencies, and stress management workshops.

Turning Military Skills into Civilian Success

When Dan transitioned from the Navy, he created a detailed action plan. He translated his military experience into civilian language, crafted a resume that highlighted his leadership and project management skills, and researched companies that valued veteran experience. Dan also built a financial safety net by budgeting and saving, utilized veteran services for career counseling, and joined support groups for additional guidance. By addressing these areas with a clear plan, Dan successfully transitioned to a fulfilling civilian career.

9

NAVIGATING FAMILY DYNAMICS AFTER SERVICE

"The strength of a nation derives from the integrity of the home."

— Confucius

The transition from military to civilian life doesn't just affect the veteran; it impacts the entire family. Every family member experiences this shift differently, and understanding these emotions is key to fostering healthy family dynamics during this critical time. This chapter provides specific, actionable strategies for managing emotional impacts, resolving conflicts, and creating a supportive home environment that strengthens your family as you navigate the transition together.

1. Understanding the Emotional Impact on Family Dynamics

. . .

Veteran's Perspective: Veterans often experience a mix of emotions, including a sense of loss over leaving a structured, purposeful environment and anxiety about finding new employment. For example, Sarah, a former Army officer, felt a profound identity crisis after service. Her frustrations over small household issues stemmed from a deeper struggle to redefine her purpose.

Family's Perspective: Spouses and children may struggle to adjust to the veteran's increased presence at home and the shift in family roles. For instance, Sarah's husband, John, had grown accustomed to making household decisions on his own during her deployments. Her sudden return disrupted this dynamic, leading to tension.

Takeaway: Recognizing that both the veteran and their family members are undergoing emotional transitions helps everyone feel heard and validated. Understanding these different perspectives is crucial to building empathy and reducing misunderstandings.

2. Strategies for Authentic Communication

Empathy in Communication: Effective communication isn't just about scheduling a weekly check-in. It's about creating an atmosphere where each family member feels comfortable sharing their true feelings. For example, Sarah and John implemented a "listening circle" at home, where each person could talk uninterrupted for five minutes about what was on their mind, from work frustrations to school challenges. This simple practice helped them gain a deeper understanding of each other's emotions.

Tailored Communication Strategies: Every family member communicates differently. While Sarah preferred face-to-face conversations, her teenage daughter felt more comfortable texting her thoughts. Adapting communication styles, such as leaving thoughtful notes or scheduling shared activities, fosters a more natural flow of dialogue.

Takeaway: Encourage open, judgment-free communication and adapt to each family member's preferred style. What works for one might not work for another, so stay flexible and patient.

3. Conflict Resolution with Compassion

Understanding Root Causes with Empathy: Conflicts often arise from unspoken emotional needs. When Sarah and John argued over small chores, they realized the real issue was their struggle to redefine their roles. Rather than focusing on surface-level disagreements, they began to ask deeper questions: "What emotions are driving this argument?" and "How can we address these feelings together?"

Customized Solutions: Personalized solutions, such as splitting household responsibilities based on individual strengths, can resolve conflicts more effectively. Sarah, for instance, took on tasks related to scheduling and logistics, where her military skills excelled, while John handled more creative tasks, like planning family outings. This division of labor allowed them both to contribute in ways that felt natural and respected.

Takeaway: Compassionate conflict resolution requires

empathy and creativity. Instead of arguing over tasks, dive deeper into the emotional needs behind the conflict and work together to develop tailored solutions.

4. Creating a Supportive and Flexible Home Environment

Adapting Routines to Emotional Needs: A rigid routine may not work for every family. When Sarah first returned home, she tried to impose a strict military-like structure on the household, which created tension. Over time, she realized that a more flexible approach, adjusting routines based on everyone's emotional and practical needs, was far more effective. Some days, she would take over morning duties so John could rest; other times, they'd swap.

Encouraging Emotional Support: Building a supportive environment means encouraging each family member to seek help when needed. Whether it's through family check-ins or being available to listen, like when Sarah's daughter opened up about her anxiety over school, creating a space where asking for support is normal can reduce feelings of isolation.

Takeaway: Flexibility is key. Be willing to adapt routines and create an atmosphere where emotional support is freely given and received.

5. Seeking Tailored External Support

Specialized Counseling: Not every counselor understands the unique challenges of military families, so finding one who specializes in this area can make a significant difference. When Sarah and John attended marriage counseling, they specifically sought out a therapist with experience in military transitions. This counselor helped them navigate the complexities of PTSD, career changes, and family reintegration.

Diverse Support Groups: Sarah also joined a veteran spouse support group where she met others dealing with similar transitions. This group gave her a safe space to express her challenges, learn coping strategies, and bond with a community that truly understood her experience.

How to Find Support:

- **Online Directories:** Websites like Military OneSource or the VA offer directories of military-friendly counselors and therapists.
- **Veteran Organizations:** Groups like the Veterans of Foreign Wars (VFW) or Blue Star Families often have local support groups for veterans and their families.
- **Financial Assistance:** Many organizations offer free or subsidized counseling for veterans and their families, so don't hesitate to ask about financial support options.

Shifting Roles: The Heart of Family After Service

. . .

I've seen it countless times—families like the Rivera's who, on paper, have everything figured out. Hector had just retired from the Marines after twenty years, coming home to his wife, Elena, and their two teenage sons, Marcos and Javier. Elena had held down the fort while Hector was away, and the boys had grown up without their father around on a daily basis. They had their own routines, their own way of managing things. But when Hector returned, things started to shift.

It's a story I hear often. What seems like a smooth homecoming quickly becomes a complicated adjustment. At first, it was small things. Hector would take over chores or suggest new routines. But soon, those small things turned into bigger issues—unspoken tension at the dinner table, small disagreements that escalated quickly. Elena found herself frustrated with decisions she used to handle on her own, while Hector tried to find his place in a household that had learned to function without him. And the boys? They didn't know how to interact with their dad, who was suddenly present for every moment.

One evening, after a particularly tense dinner, Elena reached out to me. She didn't know how to fix things but knew something had to change. Hector felt it too. "We didn't argue like this before," he told me during our first conversation. And there it was—the heart of the issue. The dynamics of the household had shifted, and no one quite knew how to adjust.

"Hector," I said, "you've been in a highly structured environment for two decades. You knew your role, you knew the mission, and you knew what success looked like. Coming home after that much time is like starting a new chapter with no roadmap."

He nodded. "I thought I'd just step back in and help, but I don't know where I fit anymore."

Elena echoed that sentiment. She loved having her husband home, but it felt like they were stepping on each other's toes. She had learned to handle things in his absence, and now it felt like they were in a power struggle over decisions that didn't need to be this hard.

I've seen this before, and I explained that their situation wasn't unique. Transitioning from military life doesn't just affect the veteran. The entire family has to adjust, especially when roles have been set over years of long-distance communication and independent decision-making. I suggested they start by acknowledging that this wasn't something they could solve overnight, but they could make small, intentional changes to make the transition smoother.

We started with a basic action plan—a framework to help them find their balance. First, I recommended they have regular check-ins, not just about logistics but about how they were feeling. "You've got to communicate like you're a team," I told them. "Because you are." Hector and Elena agreed that weekly sit-downs would help. It wasn't about solving every problem in one go; it was about making space for honest conversations.

At their first check-in, something clicked. Marcos spoke up, saying he felt like his dad was "taking over." Javier admitted he didn't know how to talk to his dad now that he was around all the time. And Hector? He realized that, in his effort to be involved, he was unintentionally disrupting the routines that had worked for years. Hearing it from the boys opened his eyes. Elena shared that she, too, felt overwhelmed by the shift in household dynamics. It was the first time they all sat down and really listened to each other.

Here's the thing—these changes are normal. You don't

come home from two decades in the military and slip seamlessly into a civilian household. It takes time, patience, and a willingness to redefine roles. Hector realized he didn't have to "fix" everything. He could be part of the household without micromanaging. Elena understood that she didn't have to carry the entire burden alone anymore, but it was okay to ease into sharing responsibilities. The boys? They needed time to adjust to having their dad fully present.

The Rivera family wasn't trying to avoid conflict; they were learning how to navigate it. They started having more regular conversations, and slowly, they found their new normal. Hector found ways to contribute that didn't upset the established routines. Elena relaxed, knowing she didn't have to be the sole decision-maker anymore. And the boys opened up more, realizing their dad wasn't trying to take over—he was just learning how to fit into a family that had operated differently while he was away.

What I always tell families is this: *transitioning takes time.* You're not going to get it right from day one, and that's okay. The key is to keep the lines of communication open and to be willing to adapt. Hector learned that just like in the military, adjusting to a new mission takes planning and execution. The mission now? Building a balanced, harmonious family life.

Action Plan:

1. **Set Up Regular Check-ins:** Create a safe space for everyone to share how they're feeling and what's working (or not) in the household. These

don't need to be long, but they should be consistent.

2. **Acknowledge the Shift in Roles:** Understand that everyone's role in the family has changed. Be patient with the process, and don't expect perfection right away.
3. **Give Each Other Grace:** This adjustment is hard on everyone—give your partner and kids the space they need to figure things out, just as you're figuring it out yourself.
4. **Seek Professional Help if Needed:** Sometimes, external support from a counselor or therapist can make all the difference. Don't hesitate to reach out if things feel too overwhelming.
5. **Be Willing to Adapt:** This new chapter is all about flexibility. If something isn't working, adjust the plan. Transitioning is fluid, and so should your approach to family life after service.

The key here is progress, not perfection. Families like the Riveras don't get it right all at once, but with time, effort, and understanding, they find their way. And you will too.

10

EDUCATION AND SKILL DEVELOPMENT

"An investment in knowledge
pays the best interest."

— Benjamin Franklin

Transitioning to civilian life often requires upgrading or acquiring new skills. This chapter will guide you through leveraging educational and skill development opportunities to advance your career. Let's break it down into clear, actionable steps.

1. Leveraging the GI Bill

Understanding Your Benefits: The GI Bill offers financial support for education and training. Here's a simplified approach to using these benefits effectively:

- **Know Your Benefits:** The GI Bill covers tuition, books, and sometimes housing. Check your eligibility and benefits details online or through a VA office.
- **Choose the Right Program:** Research programs that align with your career goals. Look for accredited institutions or vocational schools that offer training in your desired field.
- **Application Process:** Gather necessary documents, understand deadlines, and apply through the VA's online portal.

2. Exploring Vocational Training

FINDING RELEVANT PROGRAMS: Vocational training can provide practical skills for your new career. Here's how to get started:

- **Identify Programs:** Look for vocational programs that offer certifications or skills directly relevant to your career goals.
- **Evaluate Quality:** Check program accreditation, success rates, and feedback from previous participants. Ensure they have strong industry connections.
- **Utilize Support Services:** Many programs offer career counseling and job placement assistance. Take full advantage of these resources.

3. Pursuing Certification Programs

Obtaining Certifications: Certifications validate your skills and can boost your job prospects. Here's a straightforward approach:

- **Research Requirements:** Find certifications valued in your target industry. Review the coursework, exams, and experience required.
- **Prepare for Exams:** Use study guides, practice exams, and preparatory courses designed for veterans.
- **Maintain Certifications:** Stay updated on renewal requirements and continuing education

4. Bridging Skill Gaps

Addressing Skill Gaps: Identify and address any gaps in your skills with these steps:

- **Conduct a Skill Assessment:** Compare your current skills with job descriptions in your target field. Identify areas for improvement.
- **Create a Development Plan:** Outline steps to acquire necessary skills through additional training or coursework.
- **Seek Practical Experience:** Look for internships, volunteer opportunities, or part-time positions to gain hands-on experience.

5. Continuous Learning and Professional Development

Staying Current: Continuous learning is essential for career growth. Here's how to keep up:

- **Stay Informed:** Follow industry trends, join professional organizations, and attend workshops or seminars.
- **Invest in Ongoing Education:** Consider advanced degrees or specialized training to enhance your career prospects.
- **Network with Professionals:** Build a network of industry contacts to exchange knowledge and discover new opportunities.

Education and skill development are crucial for a successful transition to civilian life. By effectively using the GI Bill, exploring vocational training, pursuing certifications, bridging skill gaps, and committing to continuous learning, you can enhance your career prospects and achieve your goals. Remember, investing in your education not only prepares you for new opportunities but also sets the stage for long-term success.

A Fresh Start: Navigating Education and Skills

Meet Lisa, a Navy veteran who had spent years managing complex operations at sea. When she retired, Lisa knew she wanted to continue her work, but in a different field. She was interested in project management, a role she had admired but had never formally pursued.

Lisa started by exploring her GI Bill benefits, which offered more than just tuition coverage. With the help of an educational advisor, she learned that her benefits could cover not only a degree but also certifications that could give her a competitive edge. She chose a project management program that would give her the skills and credentials she needed to transition smoothly into her new career.

While researching, Lisa also came across vocational training programs for project management that offered hands-on experience. She enrolled in a program that provided practical training and connected her with industry professionals. The networking opportunities proved invaluable; she met mentors and colleagues who helped her understand the nuances of her new field.

One day, Lisa found herself preparing for a certification exam in project management. She used various study guides and practice exams that were tailored specifically for veterans, which helped her grasp the material and build confidence. The preparation paid off when she successfully passed the exam and earned her certification.

Lisa didn't stop there. Understanding that the job market was always evolving, she committed to continuous learning. She attended industry conferences, participated in webinars, and joined a professional organization for project managers. This commitment to staying current with industry trends and expanding her network opened doors she hadn't imagined.

With her GI Bill benefits well-utilized and her new skills in hand, Lisa secured a position as a project manager in a thriving tech company. Her journey wasn't just about changing jobs; it was about turning her military experience into a new career path, equipped with the education and skills to excel.

Lisa's story is a testament to how effective use of educational benefits, coupled with a clear plan for skill development and continuous learning, can lead to a successful transition. For any veteran like Lisa, the path to a fulfilling civilian career starts with a well-thought-out strategy and the resources available to you.

Action Steps for Education and Skill Development

1. Leveraging the GI Bill

☐ **Understand Your Benefits**

Review what the GI Bill covers, including tuition, books, and housing. Check eligibility details online or with a VA office.

☐ **Choose the Right Program**

Research and select accredited institutions or vocational schools that align with your career goals.

☐ **Apply for Benefits**

Gather necessary documents and apply through the VA's online portal, adhering to deadlines.

. . .

2. Exploring Vocational Training

☐ **Identify Relevant Programs**

Look for vocational programs offering certifications or skills related to your career goals.

☐ **Evaluate Program Quality**

Check accreditation, success rates, and feedback. Ensure programs have strong industry connections.

☐ **Utilize Support Services**

Make use of career counseling and job placement assistance provided by the program.

3. Pursuing Certification Programs

☐ **Research Certification Requirements**

Find certifications valued in your target industry and review necessary coursework, exams, and experience.

☐ **Prepare for Certification Exams**

Use study guides, practice exams, and preparatory courses designed for veterans.

☐ **Maintain Certifications**

Stay informed about renewal requirements and continue education to keep your certifications up-to-date.

4. Bridging Skill Gaps

☐ Conduct a Skill **Assessment**

Compare your skills with job descriptions in your target field to identify gaps.

☐ Create a Development **Plan**

Outline steps to acquire needed skills through training or coursework.

☐ Seek Practical **Experience**

Gain hands-on experience through internships, volunteer work, or part-time positions.

5. Continuous Learning and Professional Development

☐ Stay Informed

Follow industry trends, join professional organizations, and attend workshops or seminars.

☐ Invest **in Ongoing Education**

Consider pursuing advanced degrees or specialized training to enhance career prospects.

. . .

☐ NETWORK **with Professionals**

Build and maintain a network of industry contacts to exchange knowledge and discover new opportunities.

BY FOLLOWING THESE STEPS, you'll effectively use educational resources, bridge skill gaps, and advance your career prospects, setting yourself up for long-term success in civilian life.

11

PHYSICAL HEALTH

"Take care of your body. It's the only place you have to live."

— Jim Rohn

Maintaining and improving physical health is a cornerstone of a successful transition from military to civilian life. This chapter, aligned with the Resilience & Growth component of the S.O.A.R. framework, provides practical guidance on accessing healthcare services, managing chronic conditions, staying physically fit, and integrating mental health into your overall wellness strategy. By prioritizing your health, you build a strong foundation for resilience and well-being throughout your transition.

1. Accessing Healthcare Services

. . .

Ensuring you have access to quality healthcare is crucial for maintaining physical health. Here's how to navigate the healthcare system effectively:

- **Understanding VA Healthcare:** The Department of Veterans Affairs (VA) offers comprehensive healthcare services. Review your eligibility, understand the range of services available, and how to enroll. If VA healthcare is not available, explore private providers and community health centers.
- **Choosing the Right Provider:** Decide between VA facilities and private providers based on your specific health needs and preferences. Look for providers with experience in veteran care and those who align with your healthcare goals.
- **Managing Appointments:** Use the VA's online portal or mobile app to manage appointments, track health records, and communicate with healthcare providers. Schedule regular check-ups and preventive care to stay on top of your health.

2. Managing Chronic Conditions

Effectively managing chronic conditions is essential for maintaining quality of life. Here's how to approach this:

- **Develop a Personalized Plan:** Collaborate with your healthcare provider to create a management plan tailored to your chronic conditions. Include medication, lifestyle changes, and regular monitoring.

- **Track Symptoms and Treatment:** Use a health journal or app to record symptoms, medication use, and changes in your condition. Share this information with your healthcare provider to adjust your treatment plan as needed.
- **Seek Support:** Join support groups or online forums related to your condition. Connecting with others facing similar challenges can provide valuable insights and emotional support.

3. Staying Physically Fit

INCORPORATING regular physical activity into your routine is vital for overall health. Here's how to stay active:

- **Set Clear Fitness Goals:** Define achievable fitness goals, such as improving cardiovascular health or building strength. Use SMART criteria (Specific, Measurable, Achievable, Relevant, Time-bound) to set and track goals.
- **Create a Routine:** Develop a consistent exercise routine that includes aerobic, strength, and flexibility exercises. Aim for at least 150 minutes of moderate-intensity activity per week.
- **Find Enjoyable Activities:** Choose physical activities you enjoy to maintain motivation.

Experiment with different exercises to find what works best for you.

4. Balancing Physical Health with Daily Life

INTEGRATING health and wellness into your daily life requires balance. Here's how to manage it:

- **Incorporate Activity into Daily Tasks:** Look for ways to be active in your daily routine, such as walking or biking for short trips or taking the stairs.
- **Prioritize Self-Care:** Allocate time for activities that support physical health, including adequate sleep, stress management, and a balanced diet.
- **Seek Social Support:** Engage with family, friends, or support groups to stay motivated and accountable for your health goals.

5. Integrating Mental Health into Physical Wellness

MENTAL HEALTH IS an integral part of overall well-being. Addressing both physical and mental health ensures a comprehensive approach to wellness.

- **Recognize the Connection:** Understand how physical health impacts mental well-being and vice versa. Regular exercise can improve mood, while managing stress can enhance physical health.

- **Seek Professional Help:** If you're experiencing mental health challenges, seek support from mental health professionals. Many VA programs offer counseling and therapy services.
- **Practice Mindfulness and Stress Management:** Incorporate mindfulness practices, such as meditation or yoga, to manage stress and improve mental health.

Maintaining physical health is a crucial aspect of a successful transition to civilian life. By effectively managing healthcare services, chronic conditions, and physical fitness, while also integrating mental health strategies, you lay a strong foundation for overall well-being. Prioritizing your health not only supports your transition but also enhances your quality of life and resilience for the future.

Finding Balance: Samuel's Journey to Physical and Mental Wellness

I'd like to share a story about Samuel, a former military colleague who faced significant challenges as he transitioned from military to civilian life. Samuel, like many veterans, grappled with the dual pressures of emotional distress and physical health issues. His journey to wellness is a compelling example of how addressing both aspects can lead to a successful transition.

Samuel had always been a dedicated service member, but upon returning home, he found himself struggling with both mental and physical health. He felt overwhelmed by fatigue and anxiety, which made everyday tasks and job prospects feel daunting.

1. **Reaching Out for Comprehensive Support:** Recognizing the need for help, Samuel connected with a Veterans Service Officer (VSO), who was instrumental in guiding him through the process of accessing healthcare services. The VSO helped Samuel set up regular check-ups with a primary care physician, ensuring that he addressed his physical health needs and managed any chronic conditions effectively.

2. **Embracing Physical Fitness:** Understanding the critical role of physical fitness in overall wellness, Samuel began to incorporate exercise into his daily routine. He started with simple activities like walking, gradually progressing to more structured workouts, including strength training and yoga. This physical activity was not just about improving his health; it also provided a boost to his mental well-being, helping to alleviate anxiety and stress.

3. **Creating a Balanced Routine:** Samuel developed a routine that balanced physical exercise, healthy eating, and sufficient rest. He made a conscious effort to include nutritious foods in his diet and prioritize sleep, which contributed to his overall energy and health. This routine

gave him a sense of purpose and accomplishment, essential for navigating his transition.

4. **Leveraging Community Resources:** Samuel explored various community resources, including local fitness programs and wellness workshops. These resources offered him valuable information and support while providing opportunities for social interaction. Engaging with these programs helped him feel more connected and supported during his transition.

5. **Integrating Mental and Physical Health:** Samuel recognized that mental and physical health are deeply interconnected. Alongside his physical fitness routine, he attended mindfulness and stress management workshops. This holistic approach helped him address both his mental and physical health challenges, promoting a more balanced and resilient state of well-being.

Samuel's story is a powerful reminder of the importance of integrating physical and mental health strategies for a successful transition. By seeking comprehensive support, committing to a balanced routine, and utilizing available resources, Samuel was able to navigate his challenges and build a solid foundation for his future.

If you find yourself in a similar situation, take a page from Samuel's journey. Prioritize both your physical and mental health, seek out the right resources, and build a balanced routine. With the right support and determination, you can create a successful and fulfilling transition to civilian life.

ACTION PLAN for Maintaining Physical Health During Transition

1. **Accessing Healthcare Services**

- **Check VA Benefits:**
- Review your VA healthcare benefits online or at a local VA office.
- **Choose Providers:**
- Pick between VA and private healthcare providers based on your needs.
- **Manage Appointments:**
- Use the VA portal or app to schedule and track appointments.

Action Steps:

- Verify your benefits.
- Select healthcare providers.
- Use the VA portal for appointments.

2. **Managing Chronic Conditions**

- **Create a Plan:**
- Work with your doctor to develop a plan for managing your condition.
- **Track Your Health:**

- Use a journal or app to log symptoms and treatments.
- **Get Support:**
- Join support groups or forums for additional help.

Action Steps:

- Develop a management plan with your doctor.
- Start tracking your health.
- Find support groups.

3. Staying Physically Fit

- **Set Goals:**
- Define clear fitness goals (e.g., run a certain distance, lift weights).
- **Build a Routine:**
- Exercise regularly, aiming for 150 minutes of moderate activity per week.
- **Find Activities You Enjoy:**
- Try different exercises to see what you like.

Action Steps:

- Set fitness goals.
- Create an exercise routine.
- Explore enjoyable activities.

4. Balancing Physical Health with Daily Life

- **Stay Active Daily:**
- Include physical activity in your routine (e.g., walk or bike instead of driving).
- **Focus on Self-Care:**
- Prioritize sleep, a balanced diet, and stress management.
- **Connect with Others:**
- Engage with family and friends for support and motivation.

Action Steps:

- Incorporate activity into daily tasks.
- Maintain a self-care routine.
- Engage with your support network.

5. Integrating Mental Health with Physical Wellness

- **Understand the Link:**
- Recognize how physical health affects mental well-being.
- **Seek Help:**
- If needed, contact mental health professionals through VA programs.
- **Practice Mindfulness:**
- Use practices like meditation to manage stress.

Navigating Mental Health and Emotional Well-being

Transitioning from military to civilian life is not just a change in routine—it's a profound shift that affects your mental health and emotional well-being. This chapter is here to help you navigate these challenges with practical advice, heartfelt stories, and actionable steps to ensure you're supported every step of the way.

Recognizing Common Mental Health Challenges

Veterans face unique mental health challenges during this transition. Understanding these issues is the first step to managing them effectively.

- **PTSD**: For instance, consider Tom, a Marine who experienced vivid flashbacks and anxiety. By recognizing these as symptoms of PTSD, Tom sought specialized therapy and joined a support group, which helped him develop coping strategies.
- **Depression**: Sarah, an Army veteran, found herself feeling persistently low and disconnected. Through therapy and support from loved ones, she learned to manage her symptoms and find joy in daily activities.
- **Anxiety**: Alex, a Navy veteran, struggled with overwhelming worry about his future. By using mindfulness techniques and engaging in regular counseling, he was able to manage his anxiety more effectively.

Effective Coping Strategies

. . .

BUILDING resilience through practical strategies can help manage stress and improve your overall well-being.

- **Routine:** Establishing a daily routine, like Jack did, who started each day with exercise and scheduled time for hobbies, brought him stability and purpose.
- **Mindfulness:** Emily, who found meditation through a local veterans' group, discovered how mindfulness helped her manage stress and find inner calm.
- **Engagement:** David, a veteran who loved woodworking, found that immersing himself in this hobby provided a therapeutic escape from daily pressures.

Interactive Resource: Access the Stress Management Toolkit to explore techniques that fit your lifestyle.

BUILDING a Strong Support System

A SUPPORTIVE NETWORK can make a significant difference in managing mental health.

- **Support Groups:** Mike joined a local veteran support group, where he connected with others who understood his experiences and provided invaluable support.

- **Professional Help:** Lisa benefited from counseling tailored to veterans' needs, which helped her navigate the transition with greater confidence.
- **Family Communication:** Open dialogue with family members, as practiced by Robert, ensured mutual understanding and emotional support.

INTEGRATING Mental Health Practices into Daily Life

INCORPORATE MENTAL HEALTH practices into your routine to build long-term resilience.

- **Goal Setting:** Sarah set small, achievable goals for her mental health and celebrated her progress, which kept her motivated and positive.
- **Self-Care:** Lisa made time for activities she enjoyed, like journaling and nature walks, which were essential for maintaining her well-being.
- **Community Engagement:** David stayed connected through volunteering, which helped him maintain a sense of purpose and community.

SEEKING Long-Term Support and Growth

. . .

Ongoing support and personal growth are crucial for adapting to civilian life.

- **Evaluate and Adjust:** Regular check-ins with a counselor helped Tom adapt his strategies and stay on track with his mental health goals.
- **Personal Growth:** Emily engaged in workshops and educational programs to foster personal development and resilience.

Your mental health and emotional well-being are vital during this transition. By understanding common challenges, employing practical coping strategies, building a strong support network, and integrating mental health practices into your daily routine, you can navigate this period with resilience and confidence. Remember, support is available, and you don't have to face these challenges alone.

Finding Light in the Shadows: **Vernon's Journey to Mental Wellness**

In a quiet neighborhood, a veteran named Vernon was grappling with the challenges of transitioning from military to civilian life. The discipline and camaraderie he once knew had given way to a daunting new reality filled with anxiety and unresolved memories.

Returning home, Vernon found himself burdened by the echoes of his past service. Flashbacks and nightmares haunted him, leaving him feeling isolated from his family and friends. The transition was overwhelming, and he struggled with a deep sense of sadness and uncertainty about his future.

One evening, Vernon decided he could no longer face this struggle alone. He reached out to a friend from his service days, Mark, who had successfully navigated his own transition. Mark's words were a beacon of hope: "You don't have to go through this by yourself. There's support out there, and asking for help is a sign of strength."

Inspired by Mark's advice, Vernon joined a local veterans' support group. In this new space, he found understanding and camaraderie among others who had faced similar challenges. The group's leader, a compassionate counselor named Linda, introduced him to mindfulness practices. Although initially skeptical, Vernon began to find solace in meditation and deep-breathing exercises, which helped him manage his anxiety and regain a sense of control.

Recognizing the need for structure in his life, Vernon established a daily routine that included self-care activities and physical exercise. He took up gardening, a hobby he had always enjoyed but hadn't made time for. This new routine not only gave him a sense of accomplishment but also provided a peaceful escape from his worries.

Vernon's family also played a crucial role in his recovery. His wife, Laura, participated in family counseling with him, learning how to communicate more effectively and supportively. This shared effort helped strengthen their relationship and provided Vernon with a dependable support system.

As Vernon continued to make progress, he found value in continuous learning and personal growth. He attended workshops and engaged in community service, which not only expanded his skill set but also gave him a renewed sense of purpose and connection.

Through his journey, Vernon discovered that managing mental health and emotional well-being was an ongoing process. By embracing support, developing coping strategies, and committing to personal growth, he began to find light in the shadows and reclaim his sense of hope and direction.

Action Plan for Managing Mental Health During Transition

1. Recognizing Common Mental Health Challenges

- **Identify Symptoms:**
- Be aware of symptoms like PTSD, depression, or anxiety. Understanding these challenges is the first step in addressing them.
- **Seek Diagnosis:**
- If you recognize these symptoms in yourself, consult a mental health professional for a proper diagnosis.

Action Steps:

- Learn about common mental health challenges.
- Schedule an appointment with a mental health professional if needed.

2. Effective Coping Strategies

- **Establish a Routine:**

- Create a daily schedule that includes activities you enjoy and time for self-care.
- **Practice Mindfulness:**
- Incorporate mindfulness techniques such as meditation or deep-breathing exercises into your routine.
- **Engage in Hobbies:**
- Dedicate time to activities that provide relaxation and joy.

Action Steps:

- Set up a daily routine.
- Explore mindfulness practices.
- Engage in hobbies that bring you satisfaction.

3. Building a Strong Support System

- **Join Support Groups:**
- Find and participate in veteran support groups to connect with others who understand your experiences.
- **Seek Professional Help:**
- Consider counseling tailored to veterans' needs.
- **Communicate with Family:**
- Keep open lines of communication with family members to ensure mutual support.

Action Steps:

- Locate and join a veteran support group.

- Find a counselor experienced with veterans.
- Maintain open communication with family.

4. Integrating Mental Health Practices into Daily Life

- **Set Achievable Goals:**
- Break down mental health goals into small, manageable steps and track your progress.
- **Prioritize Self-Care:**
- Schedule regular self-care activities such as journaling, exercise, or spending time in nature.
- **Engage with the Community:**
- Participate in community service or events to stay connected and purposeful.

Action Steps:

- Set small, achievable mental health goals.
- Incorporate self-care activities into your routine.
- Get involved in community activities.

5. Seeking Long-Term Support and Growth

- **Regular Check-Ins:**
- Have periodic evaluations with a counselor to adjust strategies as needed.
- **Focus on Personal Growth:**
- Attend workshops or educational programs to foster ongoing development.

Action Steps:

- Schedule regular counseling sessions.
- Pursue opportunities for personal growth.

Use this action plan to effectively manage your mental health during your transition. Focus on recognizing challenges, employing coping strategies, building support systems, integrating mental health practices into daily life, and seeking long-term support. With these steps, you'll build a resilient and supportive foundation for your well-being.

12

RESILIENCE & GROWTH

"The oak fought the wind and was broken, the willow bent when it must and survived."

— Robert Jordan

The transition from military to civilian life is more than a change in occupation—it's a profound shift that requires resilience and a commitment to personal growth. This chapter explores the final component of the S.O.A.R. framework—Resilience & Growth—and provides practical strategies for adapting to this new phase of life with strength and optimism.

1. Embracing Change

Change can be challenging, but it is also an opportunity for growth. Embracing change involves:

- **Acknowledging Your Feelings:** It's natural to feel uncertain or anxious about change. Recognize these feelings as part of the transition process and give yourself permission to experience them.
- **Staying Open-Minded:** Approach new experiences and opportunities with an open mind. Flexibility and adaptability are key to navigating unfamiliar terrain.
- **Setting Realistic Expectations:** Understand that adjustment takes time. Set achievable goals and be patient with yourself as you adapt to new circumstances.

Personal Story: Tom, a veteran transitioning to a new career, initially struggled with the changes in his routine. By staying open-minded and setting small, achievable goals, he gradually embraced his new role and found success.

2. Committing to Lifelong Learning

Continuing education and skill development are vital for personal and professional growth:

- **Pursue Further Education:** Take advantage of educational programs, certifications, and workshops to enhance your skills and knowledge. The GI Bill and other benefits can support your educational goals.
- **Explore New Interests:** Engage in activities or hobbies that interest you. Lifelong learning can include non-academic pursuits, such as

learning a new skill or exploring a passion project.

- **Seek Mentorship and Guidance:** Find mentors or advisors who can provide valuable insights and advice as you navigate new career paths or personal goals.

3. Celebrating Milestones

Recognizing and celebrating your achievements, no matter how small, can boost motivation and morale:

- **Acknowledge Your Progress:** Regularly review your goals and celebrate the progress you've made. Recognizing small victories can build confidence and keep you motivated.
- **Set Up Reward Systems:** Create a system of rewards for reaching milestones. This can be as simple as treating yourself to something you enjoy or taking time to reflect on your achievements.
- **Share Successes with Others:** Celebrate your accomplishments with friends, family, or mentors. Sharing your successes can strengthen your support network and provide additional encouragement.

4. Building and Sustaining Resilience

. . .

Resilience is the ability to bounce back from challenges and setbacks. To build and sustain resilience:

- **Develop Coping Strategies:** Identify effective coping mechanisms for dealing with stress and adversity. Techniques such as mindfulness, exercise, and healthy eating can support mental and physical resilience.
- **Maintain a Positive Outlook:** Cultivate a positive mindset by focusing on what you can control and seeking opportunities for growth. Surround yourself with supportive individuals who encourage and uplift you.
- **Reflect and Adjust:** Regularly reflect on your experiences and adjust your strategies as needed. Resilience involves ongoing learning and adaptation.

5. Cultivating a Growth Mindset

A growth mindset involves believing in your ability to develop and improve through effort and learning:

- **Embrace Challenges:** View challenges as opportunities for growth rather than obstacles. Embracing difficult situations can lead to personal development and increased confidence.

- **Learn from Feedback:** Accept constructive criticism as a chance to learn and improve. Feedback can provide valuable insights and help you refine your skills and approach.
- **Celebrate Learning:** Focus on the learning process rather than just the end result. Appreciate the effort and progress you make along the way.

Resilience and growth are integral to a successful transition from military to civilian life. By embracing change, committing to lifelong learning, celebrating milestones, building resilience, and cultivating a growth mindset, you can navigate this transition with confidence and adaptability. Remember, every step you take toward personal and professional development contributes to your overall success and well-being.

Growing Strong: Vernon's Path to Resilience and Growth

Vernon, a veteran who had served in the military for over a decade, was stepping into a new chapter of his life. The transition from active duty to civilian life was both thrilling and daunting. He was excited about new opportunities but found himself struggling with uncertainty and self-doubt.

The first few months were especially tough for Vernon. He faced a barrage of new challenges, from adjusting to a new career to navigating daily life without the structure of military routines. The anxiety and stress of these changes felt overwhelming.

Recognizing that he needed help, Vernon reached out to his old squad mate, Greg, who had gone through a similar transition. Greg shared his own journey of embracing change, emphasizing the importance of acknowledging his feelings and staying open-minded. “It’s okay to feel anxious,” Greg advised. “But remember, every challenge is an opportunity to grow.”

Inspired by Greg’s advice, Vernon decided to take concrete steps towards his personal growth. He enrolled in a vocational training program to learn new skills relevant to his career interests. Each step forward, whether it was completing a certification or attending a workshop, gave him a sense of accomplishment. He set up a system to celebrate each milestone, treating himself to small rewards like a favorite meal or a day out with his family.

To build resilience, Vernon adopted a daily routine that included mindfulness practices and regular exercise. He found solace in journaling about his experiences and setting small, achievable goals. This approach not only helped him manage stress but also allowed him to track his progress and reflect on his growth.

Building a support network was also crucial. Vernon joined a local veterans’ support group where he found a community of individuals who understood his journey. Through shared experiences and advice, he learned effective coping strategies and felt less isolated.

One of the most transformative aspects of Vernon’s journey was cultivating a growth mindset. He embraced challenges as learning opportunities and welcomed feedback from mentors. This mindset shift helped him overcome obstacles and find new paths for growth.

Months later, Vernon looked back at his transition with pride. He had not only adapted to his new role but had also grown personally and professionally. By embracing change, committing to lifelong learning, and celebrating his milestones, Vernon navigated his transition with resilience and optimism.

His story stands as a testament to the power of resilience and growth, showing that with the right mindset and support, the transition from military to civilian life can lead to profound personal development and success.

~

ACTION PLAN for Building Resilience and Embracing Growth

1. Embracing Change

- **Acknowledge Your Feelings:**
- Recognize and accept your emotions about the transition. It's normal to feel uncertain or anxious.
- **Stay Open-Minded:**
- Approach new experiences with curiosity and flexibility. Be willing to adapt to new situations.
- **Set Realistic Expectations:**
- Understand that adapting takes time. Set achievable short-term goals and be patient with your progress.

Action Steps:

- Reflect on your feelings and discuss them with a trusted friend or counselor.

- List new experiences you'll encounter and approach them with an open attitude.
- Set small, manageable goals for your transition and review your progress regularly.

2. Committing to Lifelong Learning

- **Pursue Further Education:**
- Explore educational programs, certifications, and workshops that align with your career and personal interests.
- **Explore New Interests:**
- Engage in hobbies or activities you're passionate about. Lifelong learning includes non-academic pursuits too.
- **Seek Mentorship and Guidance:**
- Find mentors or advisors who can offer insights and support in your new career or personal goals.

Action Steps:

- Research and enroll in relevant educational programs or workshops.
- Identify and start new hobbies or interests.
- Reach out to potential mentors for guidance and support.

3. Celebrating Milestones

- **Acknowledge Your Progress:**
- Regularly review and recognize your achievements, both big and small.
- **Set Up Reward Systems:**
- Create a reward system for achieving milestones, whether it's a treat or a break to reflect on your success.
- **Share Successes with Others:**
- Share your achievements with friends, family, or mentors to strengthen your support network.

Action Steps:

- Keep a journal of your progress and milestones.
- Plan small rewards for achieving your goals.
- Share your successes in conversations or social media to engage your support network.

4. Building and Sustaining Resilience

- **Develop Coping Strategies:** Identify and practice coping mechanisms such as mindfulness, regular exercise, and healthy eating.
- **Maintain a Positive Outlook:** Focus on what you can control and seek growth opportunities. Surround yourself with positive influences.
- **Reflect and Adjust:** Regularly assess your coping strategies and adjust them as needed. Resilience is about continuous learning and adaptation.

Action Steps:

- Implement coping strategies into your daily routine.
- Maintain a positive mindset by focusing on achievable goals.
- Schedule periodic self-reflections to adjust your strategies and goals.

5. Cultivating a Growth Mindset

- **Embrace Challenges:**
- View challenges as opportunities for growth and learning.
- **Learn from Feedback:**
- Accept constructive criticism as a tool for improvement. Use feedback to refine your skills and approach.
- **Celebrate Learning:**
- Focus on the learning process and the progress you make, not just the final outcome.

Action Steps:

- Approach new challenges with a mindset of learning and growth.
- Seek and reflect on feedback from mentors or colleagues.
- Celebrate the progress and skills you gain through learning.

To BUILD resilience and foster growth, acknowledge your feelings, stay open to change, and set realistic expectations. Commit to lifelong learning through education and new interests, celebrate your milestones, and develop coping strategies. Cultivate a growth mindset by embracing challenges, learning from feedback, and focusing on the learning process. By following these steps, you'll strengthen your resilience and navigate your transition with confidence and optimism.

13

SETTING LONG-TERM GOALS

"Setting goals is the first step in turning the invisible into the visible."

— Tony Robbins

Transitioning from military to civilian life opens up new horizons and possibilities. To ensure sustained success, it's essential to focus on setting and achieving long-term goals. This chapter, building on the Resilience & Growth aspect of the S.O.A.R. framework, provides a comprehensive guide to navigating your future with clarity and purpose.

1. Defining Your Long-Term Goals

Setting clear and achievable long-term goals is fundamental to success:

- **Identify Your Vision:** Start by envisioning where you want to be in 5, 10, or even 20 years. Think about your career aspirations, personal dreams, and financial objectives.
- **Set Specific and Measurable Goals:** Break down your vision into specific, measurable goals. For instance, if you aim to advance in your career, define what positions or roles you aspire to and the steps needed to get there.
- **Create a Roadmap:** Develop a detailed plan outlining the steps and milestones required to achieve your long-term goals. Include timelines and resources needed.o define your long-term goals, set milestones, and create a detailed action plan.

2. Exploring Career Advancement Opportunities

Career growth is a critical component of long-term success:

- **Pursue Professional Development:** Engage in continuous learning through certifications, advanced degrees, or specialized training relevant to your field. This will keep you competitive and open doors to new opportunities.
- **Seek Mentorship:** Connect with mentors who can provide guidance, support, and advice on career advancement. Their experience can offer

valuable insights and help you navigate your career path.

- **Build a Professional Network:** Networking is crucial for career growth. Attend industry events, join professional organizations, and connect with peers to expand your network and discover new opportunities.

3. Considering Entrepreneurship Opportunities

For those interested in starting their own business:

- **Explore Business Ideas:** Assess your skills, interests, and market needs to identify viable business ideas. Consider how your military experience can contribute to your entrepreneurial venture.
- **Develop a Business Plan:** Create a comprehensive business plan outlining your vision, target market, financial projections, and operational strategies. A solid plan is essential for securing funding and guiding your business.
- **Utilize Resources:** Take advantage of resources available for veterans, such as business grants, mentorship programs, and entrepreneurial training.

4. Implementing Financial Growth Strategies

Financial stability is crucial for long-term success:

- **Create a Financial Plan:** Develop a comprehensive financial plan that includes budgeting, saving, investing, and managing debt. A well-structured plan will help you achieve your financial goals and secure your future.
- **Explore Investment Opportunities:** Look into investment options such as stocks, real estate, or retirement accounts to grow your wealth. Consult with a financial advisor to make informed decisions.
- **Monitor and Adjust:** Regularly review your financial plan and adjust as needed based on changes in your goals or circumstances.

5. Celebrating Achievements and Reflecting on Progress

Acknowledging and celebrating your successes reinforces motivation and commitment:

- **Celebrate Milestones:** Recognize and celebrate each achievement along the way. Celebrating milestones, no matter how small, boosts morale and keeps you motivated.
- **Reflect on Your Journey:** Regularly review your progress, reflect on your achievements, and adjust your goals as needed. Reflecting on your journey helps you stay focused and inspired.

SETTING and achieving long-term goals requires vision, planning, and persistence. By defining your goals, exploring career and entrepreneurial opportunities, implementing financial growth strategies, and celebrating your successes, you can ensure sustained success in your civilian life. Embrace the journey with courage and determination, knowing that every step you take brings you closer to achieving your dreams.

NAVIGATING NEW HORIZONS: Vernon's Journey to Long-Term Success

AFTER LEAVING THE MILITARY, Vernon was excited about the endless possibilities of civilian life but felt overwhelmed by the vastness of his goals. He knew he wanted to make a significant impact, but figuring out how to translate his military experience into a successful civilian career was daunting.

Vernon started by defining his long-term goals. He envisioned himself becoming a leader in the non-profit sector, focusing on veteran services. To turn this vision into reality, he broke it down into specific, measurable goals: gaining relevant certifications, networking with industry professionals, and securing a leadership position within five years.

Creating a detailed roadmap was his next step. Vernon set short-term goals, like completing a leadership training program within six months and attending networking events regularly. He used a Goal-Setting Worksheet to map out these steps, setting clear deadlines and identifying resources needed.

One challenge Vernon faced was balancing his professional development with personal commitments. He found that enrolling in a time management workshop and using a Career Advancement Planner helped him stay organized and focused. By setting realistic expectations and adjusting his plan as needed, Vernon managed to make steady progress.

Inspired by his own experiences, Vernon also explored entrepreneurship. He realized that his passion for helping others could be channeled into a social enterprise. He developed a detailed Business Plan, incorporating his military background and leadership skills to craft a business idea focused on providing support services for veterans.

Financial planning was another critical aspect of Vernon's journey. He created a comprehensive Financial Plan that included budgeting for his business startup and investing in his personal development. By regularly reviewing his Financial Growth Tracker, he stayed on top of his investments and adjusted his strategies to align with his evolving goals.

As Vernon achieved milestones—like completing his leadership certification or securing his first major funding for his business—he made it a point to celebrate these victories. Whether it was treating himself to a special dinner or sharing his success with friends and family, these celebrations kept him motivated and focused.

Reflecting on his journey, Vernon found that his resilience and commitment to growth were key to his success. He made it a habit to regularly review his progress and adjust his goals as needed. This reflective practice not only helped him stay on track but also inspired him to continue pursuing new opportunities.

Vernon's story illustrates the power of setting and

achieving long-term goals. By defining his vision, breaking it down into actionable steps, and celebrating his progress, he navigated his transition with clarity and purpose. His journey serves as a reminder that with determination and a well-structured plan, the transition to civilian life can lead to profound personal and professional success.

Action Plan for Setting and Achieving Long-Term Goals

1. Defining Your Long-Term Goals

- **Identify Your Vision:**
- Envision where you want to be in the next 5, 10, or 20 years, considering career, personal dreams, and financial objectives.
- **Set Specific and Measurable Goals:**
- Break down your vision into clear, specific goals. For example, if you want to advance in your career, define the positions or roles you aim for and the steps needed to reach them.
- **Create a Roadmap:**
- Develop a detailed plan with steps, milestones, timelines, and resources needed to achieve your long-term goals.

Action Steps:

- Write down your long-term vision and goals.
- Break down your vision into smaller, measurable goals.

- Create a detailed roadmap with timelines and resources. Use a Goal-Setting Worksheet to organize your plan.

2. Exploring Career Advancement Opportunities

- **Pursue Professional Development:**
- Engage in continuous learning through certifications, advanced degrees, or specialized training relevant to your field.
- **Seek Mentorship:**
- Connect with mentors who can provide guidance and support on career advancement.
- **Build a Professional Network:**
- Attend industry events, join professional organizations, and network with peers to discover new opportunities.

Action Steps:

- Research and enroll in relevant educational programs or certifications.
- Identify and reach out to potential mentors.
- Attend networking events and join professional groups in your field.

3. Considering Entrepreneurship Opportunities

- **Explore Business Ideas:**

- Assess your skills, interests, and market needs to identify viable business ideas.
- **Develop a Business Plan:**
- Create a comprehensive plan outlining your vision, target market, financial projections, and operational strategies.
- **Utilize Resources:**
- Take advantage of resources available for veterans, such as business grants, mentorship programs, and entrepreneurial training.

Action Steps:

- Brainstorm and evaluate potential business ideas.
- Develop a detailed business plan.
- Explore and apply for veteran-specific resources and support programs.

4. Implementing Financial Growth Strategies

- **Create a Financial Plan:**
- Develop a plan that includes budgeting, saving, investing, and managing debt to achieve your financial goals.
- **Explore Investment Opportunities:**
- Look into investment options like stocks, real estate, or retirement accounts. Consult with a financial advisor for informed decisions.
- **Monitor and Adjust:**

- Regularly review your financial plan and adjust based on changes in your goals or circumstances.

Action Steps:

- Create and implement a comprehensive financial plan.
- Research investment options and consult with a financial advisor.
- Review and adjust your financial plan regularly.

5. Celebrating Achievements and Reflecting on Progress

- **Celebrate Milestones:**
- Recognize and celebrate each achievement to boost motivation and morale.
- **Reflect on Your Journey:**
- Regularly review your progress and reflect on your achievements. Adjust goals as needed to stay focused and inspired.

Action Steps:

- Set up a system to track and celebrate milestones.
- Schedule regular reflection sessions to review progress and adjust goals.
- Share your achievements with friends and family for additional support.

Summary: Setting and achieving long-term goals requires clear vision, planning, and persistence. By defining your goals, exploring career and entrepreneurial opportunities, implementing financial strategies, and celebrating your successes, you can navigate your transition with purpose and confidence. Embrace the journey with determination, and every step will bring you closer to achieving your dreams.

14

STAYING CONNECTED WITH THE VETERAN COMMUNITY

"We must find time to stop and thank the people who make a difference in our lives."

— John F. Kennedy

Staying connected with the veteran community is a vital component of navigating the transition to civilian life. Building and maintaining these connections not only fosters a sense of belonging but also provides ongoing support and opportunities for personal and professional growth. This chapter focuses on the importance of engagement and offers practical ways to stay involved with the veteran community, aligning with the Resilience & Growth step of the S.O.A.R. framework.

1. The Value of Community Connections

. . .

MAINTAINING relationships within the veteran community provides numerous benefits:

- **Support Network:** Fellow veterans can offer unique insights and support based on shared experiences. This network can be a valuable resource for advice, encouragement, and camaraderie.
- **Sense of Belonging:** Staying connected helps preserve a sense of belonging and purpose, which is crucial for overall well-being during the transition to civilian life.
- **Opportunities for Growth:** Engaging with the community can open doors to mentorship, career opportunities, and personal development.

2. Getting Involved with Veteran Organizations

MANY ORGANIZATIONS OFFER opportunities to connect and contribute:

- **Join Veteran Service Organizations (VSOs):** Organizations such as the VFW (Veterans of Foreign Wars), American Legion, and DAV (Disabled American Veterans) provide support, advocacy, and social activities for veterans. Joining these groups can help you stay connected and informed.
- **Participate in Local Chapters:** Many VSOs have local chapters that host events, meetings, and

volunteer opportunities. Getting involved in your local chapter can strengthen community ties and expand your network.

- **Attend Events and Conferences:** Participate in veteran-focused events, workshops, and conferences. These gatherings offer opportunities for networking, learning, and sharing experiences.

3. Engaging in Mentorship Programs

MENTORSHIP CAN BE a powerful way to give back and grow:

- **Become a Mentor:** Use your experience to guide and support fellow veterans who are navigating their own transitions. Mentoring can be a rewarding way to share your knowledge and make a positive impact.
- **Seek a Mentor:** If you're in need of guidance, seek out a mentor who can provide valuable insights and support. Look for individuals with experience in areas you're interested in or those who have successfully navigated similar transitions.
- **Participate in Mentorship Programs:** Many organizations offer structured mentorship programs that match veterans with mentors based on their needs and goals.

4. Advocating for Veteran Issues

Engagement in advocacy can amplify your voice and effect positive change:

- **Get Involved in Advocacy Initiatives:** Join or support advocacy groups that work on issues affecting veterans, such as policy changes, benefits improvements, and community support.
- **Raise Awareness:** Use your platform to raise awareness about veteran issues and promote initiatives that support the community.
- **Participate in Legislative Efforts:** Engage in legislative advocacy by participating in campaigns, contacting representatives, and attending town hall meetings.

5. Maintaining Connections

Regularly nurturing your connections is key to staying engaged:

- **Stay in Touch:** Make an effort to stay in contact with fellow veterans through calls, emails, or social media. Regular communication helps maintain and strengthen relationships.
- **Offer Support:** Be proactive in offering support to others in the community. Acts of kindness and

assistance reinforce bonds and build a supportive network.
- **Participate in Social Activities:** Engage in social activities and gatherings with veteran friends and organizations. Shared experiences and interactions contribute to a strong sense of community.

Staying connected with the veteran community is essential for continued growth, support, and well-being. By engaging with veteran organizations, participating in mentorship programs, advocating for veteran issues, and nurturing your relationships, you can maintain a strong sense of belonging and purpose. Embrace the community that has supported you and contribute to its strength and resilience.

Vera's Journey of Community and Support

When Vera transitioned from military to civilian life, she felt a mix of excitement and uncertainty. The familiar camaraderie of her military days was replaced by a sea of new faces and unfamiliar routines. Despite her achievements, she realized something crucial was missing—a connection to the veteran community that had been her second family for years.

One day, while browsing online, Vera stumbled upon a local Veterans of Foreign Wars (VFW) post. Curious and eager to reconnect with her roots, she decided to attend a meeting. The first gathering was a bit overwhelming, but Vera noticed the warmth and shared experiences among the veterans. They spoke about challenges, victories, and everything in between—stories that resonated deeply with her.

Vera soon joined the VFW and became an active member. She took on responsibilities such as organizing community events and participating in outreach programs. Through these activities, she not only rebuilt her network but also found a renewed sense of purpose. Her involvement in local events gave her the opportunity to contribute to causes she cared about and connect with veterans who understood her experiences.

During her time with the VFW, Vera also discovered the value of mentorship. She had the chance to mentor newer veterans who were navigating their own transitions. Sharing her knowledge and supporting others gave her a profound sense of fulfillment. It was rewarding to see her mentees thrive, knowing that she had played a role in their journey.

Her commitment to the veteran community extended beyond her local post. Vera became involved in advocacy efforts, working to improve veteran benefits and raise awareness about issues affecting her peers. Her passion for these causes amplified her connection to the community and allowed her to contribute to meaningful change.

Maintaining these connections was crucial for Vera. She made it a point to stay in touch with her veteran friends, whether through phone calls, social media, or attending local gatherings. Offering support and participating in social activities helped her reinforce these bonds and maintain a strong network.

Vera's story exemplifies the power of staying connected with the veteran community. By joining organizations, engaging in mentorship, and participating in advocacy, she created a network of support that enriched her life and contributed to her growth. Her journey highlights the importance of maintaining these connections and embracing the community that has supported her through the transition to civilian life.

Action Plan for Staying Connected with the Veteran Community

1. The Value of Community Connections

- **Support Network:**
- Engage with fellow veterans to leverage their unique insights and support.
- **Sense of Belonging:**
- Maintain a sense of purpose and belonging by staying active within the veteran community.
- **Opportunities for Growth:**
- Seek out mentorship and career opportunities through community involvement.

Action Steps:

- Identify veteran networks and support groups relevant to your interests and needs.
- Reach out to fellow veterans and participate in community activities.

2. Getting Involved with Veteran Organizations

- **Join Veteran Service Organizations (VSOs):**
- Consider organizations like the VFW, American Legion, and DAV for support and social activities.
- **Participate in Local Chapters:**

- Get involved with local chapters to strengthen community ties and expand your network.
- **Attend Events and Conferences:**
- Engage in veteran-focused events and conferences for networking and learning.

Action Steps:

- Research and join relevant VSOs.
- Attend local chapter meetings and volunteer for events.
- Participate in veteran-focused workshops and conferences.

3. Engaging in Mentorship Programs

- **Become a Mentor:**
- Offer guidance and support to fellow veterans navigating their transition.
- **Seek a Mentor:**
- Find a mentor who can provide insights and advice on your career and personal goals.
- **Participate in Mentorship Programs:**
- Engage in structured mentorship programs that align with your needs and goals.

Action Steps:

- Identify opportunities to mentor others within veteran organizations.
- Search for and reach out to potential mentors.

- Join a formal mentorship program to connect with experienced mentors.

4. Advocating for Veteran Issues

- **Get Involved in Advocacy Initiatives:**
- Support groups working on veteran issues such as policy changes and community support.
- **Raise Awareness:**
- Use your platform to promote veteran issues and initiatives.
- **Participate in Legislative Efforts:**
- Engage in advocacy campaigns, contact representatives, and attend town hall meetings.

Action Steps:

- Research advocacy groups and join their initiatives.
- Share information about veteran issues through social media or community platforms.
- Participate in advocacy events and legislative campaigns.

5. Maintaining Connections

- **Stay in Touch:**
- Keep regular contact with veteran friends through calls, emails, or social media.

- **Offer Support:**
- Proactively offer help and support to others in the veteran community.
- **Participate in Social Activities:**
- Engage in social gatherings and activities to strengthen relationships.

Action Steps:

- Set reminders to check in with veteran friends and acquaintances.
- Look for opportunities to support fellow veterans in need.
- Attend social events and gatherings organized by veteran groups.

SUMMARY: Staying connected with the veteran community is vital for continued support and personal growth. By getting involved with veteran organizations, participating in mentorship programs, advocating for veteran issues, and maintaining connections, you can build a strong support network and enrich your transition to civilian life. Embrace the community that has been a cornerstone of your journey and contribute to its strength and resilience.

15

RESILIENCE AND ADAPTABILITY

"It's not whether you get knocked down,
it's whether you get up."

— Vince Lombardi

As we conclude the main content of our guide, we turn our focus to one of the most crucial aspects of the transition to civilian life: resilience and adaptability. The journey from military to civilian life is marked by challenges and changes, but with a strong mindset and the right strategies, veterans and their families can overcome obstacles and thrive in their new environment.

1. Embracing Resilience

. . .

RESILIENCE IS the ability to bounce back from adversity, learn from setbacks, and continue moving forward. It's a critical skill for navigating the complexities of transitioning to civilian life.

- **Acknowledge Challenges:** Recognize that the transition may come with difficulties and that it's okay to experience setbacks. Embracing these challenges as part of the process can help you build resilience.
- **Learn from Setbacks:** Use challenges as opportunities to grow. Reflect on what went wrong, identify lessons learned, and apply these insights to future endeavors.
- **Stay Positive:** Maintaining a positive attitude can help you remain focused and motivated. Celebrate small victories and remind yourself of your strengths and achievements.

2. Adapting to Change

ADAPTABILITY IS essential for navigating the evolving landscape of civilian life. Being flexible and open to change can help you manage transitions more effectively.

- **Embrace New Experiences:** Be open to trying new things and stepping out of your comfort zone. Embracing new experiences can lead to personal growth and new opportunities.

- **Adjust Your Goals:** As circumstances change, be willing to adjust your goals and plans. Flexibility can help you stay aligned with your evolving needs and aspirations.
- **Seek Feedback:** Regularly seek feedback from mentors, peers, and support networks. Their insights can help you adapt and improve your approach to various challenges.

3. Building a Support Network

A STRONG SUPPORT network can provide encouragement, advice, and practical assistance as you navigate your transition.

- **Connect with Mentors:** Seek out mentors who can provide guidance and support. Their experience and advice can help you navigate challenges and achieve your goals.
- **Engage with Support Groups:** Participate in support groups and veteran organizations. These groups offer a sense of community and resources for managing the transition.
- **Utilize Professional Resources:** Take advantage of career counseling, mental health services, and other professional resources designed to support veterans.

4. Practical Tips for Staying Motivated

- **Set Realistic Goals:** Establish clear, achievable goals to maintain focus and motivation. Break larger goals into smaller, manageable tasks.
- **Develop a Routine:** Create a daily routine that incorporates healthy habits, work, and personal time. A structured routine can help you stay organized and motivated.
- **Celebrate Milestones:** Recognize and celebrate your achievements, no matter how small. Celebrating milestones can boost your morale and motivation.

Resilience and adaptability are key to a successful transition to civilian life. By embracing challenges, adapting to change, building a strong support network, and staying motivated, you can overcome obstacles and achieve long-term success. Remember, your journey is unique, and with the right mindset and resources, you can navigate this transition with confidence and strength.

Resilience and Adaptability: Liam's Path to New Beginnings

Liam, a decorated veteran, faced a turning point as he transitioned from his military career to civilian life. The discipline and structure he was accustomed to were suddenly replaced by a world that felt unpredictable and unfamiliar. His transition was marked by several challenges, including finding a new career and adjusting to a different pace of life.

At first, Liam struggled with the change. He experienced job rejections and felt the weight of uncertainty. However, he remembered a crucial lesson from his military days: resilience. Instead of letting setbacks define his journey, Liam chose to view them as stepping stones. He took time to acknowledge his challenges, reflecting on each setback as an opportunity to learn and grow.

Liam's adaptability played a vital role in his transition. He embraced new experiences with an open mind. When a friend suggested he try a career in tech, something he had never considered before, Liam took the plunge. He enrolled in coding classes and began networking with industry professionals. His willingness to step out of his comfort zone opened doors he hadn't anticipated.

Throughout his journey, Liam built a strong support network. He connected with a mentor who had successfully navigated a similar transition. Their guidance and encouragement were instrumental in helping him refine his approach and stay focused on his goals. Additionally, Liam joined a veteran support group where he found camaraderie and practical advice from peers who understood his experiences.

Liam's story is a testament to the power of resilience and adaptability. His ability to bounce back from challenges, adapt to new circumstances, and seek support when needed led to a successful career transition. He found fulfillment in his new tech role, and his experiences inspired others in his support group to tackle their own transitions with renewed confidence.

By staying motivated and celebrating small victories along the way, Liam demonstrated how resilience and adaptability are crucial for navigating the complexities of civilian life. His journey highlights that, with the right mindset and support, overcoming obstacles and achieving success is not only possible but also profoundly rewarding.

Action Plan for Building Resilience and Adaptability

1. Embracing Resilience

- **Acknowledge Challenges:**
- Recognize that setbacks are part of the transition process.
- Allow yourself to experience and address difficulties without self-judgment.
- **Learn from Setbacks:**
- Reflect on challenges to understand what went wrong.
- Identify lessons learned and apply these insights to future efforts.
- **Stay Positive:**
- Maintain a positive outlook even in the face of adversity.
- Celebrate small victories and remind yourself of your strengths and achievements.

Action Steps:

- Keep a journal to document challenges, lessons learned, and progress.
- Set aside time for self-reflection to acknowledge and address difficulties.

- Create a list of strengths and past successes to review when feeling discouraged.

2. Adapting to Change

- **Embrace New Experiences:**
- Be open to trying new activities and stepping out of your comfort zone.
- Seek opportunities for personal and professional growth.
- **Adjust Your Goals:**
- Reevaluate and modify your goals as circumstances change.
- Stay flexible and adapt your plans to align with your evolving needs and aspirations.
- **Seek Feedback:**
- Regularly seek feedback from mentors, peers, and support networks.
- Use feedback to refine your approach and make necessary adjustments.

Action Steps:

- Explore new hobbies or career opportunities to broaden your experience.
- Review and adjust your goals quarterly to ensure they remain relevant.
- Schedule regular check-ins with mentors or peers to gather feedback.

3. Building a Support Network

- **Connect with Mentors:**
- Identify and reach out to mentors who can provide guidance and support.
- Leverage their experience to navigate challenges and achieve your goals.
- **Engage with Support Groups:**
- Join support groups and veteran organizations for a sense of community.
- Participate in group activities and discussions to build relationships.
- **Utilize Professional Resources:**
- Take advantage of career counseling, mental health services, and other professional resources.
- Seek out resources designed specifically for veterans to support your transition.

Action Steps:

- Create a list of potential mentors and contact them for support.
- Research and join local veteran support groups or organizations.
- Identify and schedule appointments with career counselors or mental health professionals.

5. Practical Tips for Staying Motivated

- **Set Realistic Goals:**

- Break down larger goals into smaller, manageable tasks.
- Regularly review and adjust goals to ensure they are achievable.
- **Develop a Routine:**
- Create a structured daily routine that includes work, personal time, and healthy habits.
- Maintain consistency to stay organized and motivated.
- **Celebrate Milestones:**
- Acknowledge and celebrate each achievement, no matter how small.
- Use celebrations to boost morale and reinforce motivation.

Resilience and adaptability are critical for successfully navigating the transition to civilian life. By embracing challenges, adapting to change, building a strong support network, and staying motivated, you can overcome obstacles and achieve long-term success. Approach your journey with a positive mindset and utilize the resources and strategies outlined in this chapter to enhance your resilience and adaptability.

100 ORGANIZATIONS VETERANS SHOULD KNOW

Organizations Veterans Should Be Aware Of For Transition Assistance

1. **U.S. Department of Veterans Affairs (VA)** - The VA provides comprehensive healthcare, benefits, and support services to veterans and their families, including disability compensation, pension programs, and healthcare services. www.va.gov
2. **Veterans Employment and Training Service (VETS)** - VETS, part of the U.S. Department of Labor, helps veterans find jobs by providing resources for employment and career development. www.dol.gov/agencies/vets
3. **Hire Heroes USA** - A nonprofit organization offering free job search assistance, career counseling, and resume workshops to veterans and military spouses. www.hireheroesusa.org

4. **Wounded Warrior Project** - Wounded Warrior Project supports veterans with physical and mental health recovery, career development, and connection to community resources. www.woundedwarriorproject.org
5. **The Mission Continues** - Focused on empowering veterans to continue serving their communities through leadership programs and volunteering opportunities. www.missioncontinues.org
6. **Team Rubicon** - A veteran-led disaster response organization that offers veterans a new sense of purpose through humanitarian service projects. www.teamrubiconusa.org
7. **Veterans of Foreign Wars (VFW)** - A nonprofit organization that provides support, advocacy, and community for veterans through various assistance programs. www.vfw.org
8. **American Legion** - The American Legion helps veterans and their families with benefits, education, and job placement services. www.legion.org
9. **Student Veterans of America (SVA)** - A nonprofit organization that connects student veterans with resources to help them succeed in higher education and in their careers. www.studentveterans.org
10. **Tragedy Assistance Program for Survivors (TAPS)** - TAPS offers compassionate care and support to families grieving the loss of a military loved one. www.taps.org
11. **Disabled American Veterans (DAV)** - DAV provides a variety of services to disabled

veterans, including benefits assistance, employment resources, and advocacy. www.dav.org

12. **GI Bill** - The GI Bill provides educational benefits to veterans, including tuition, housing, and supplies assistance for college and training programs. www.benefits.va.gov/gibill
13. **Patriot Boot Camp** - Offers entrepreneurial education, mentorship, and community for veterans interested in starting their own businesses. www.patriotbootcamp.org
14. **USO (United Service Organizations)** - The USO offers transition assistance, mentoring, networking, and employment resources for veterans and military families. www.uso.org
15. **Veterans Business Outreach Center (VBOC)** - Provides veterans with entrepreneurial development services such as business training, counseling, and mentoring. www.sba.gov/vboc
16. **Veterans Crisis Line** - A confidential, toll-free hotline for veterans and their families to connect with mental health professionals in times of crisis. www.veteranscrisisline.net
17. **The Veterans Consortium Pro Bono Program** - Provides free legal assistance to veterans and their families who need help with VA disability appeals. www.vetsprobono.org
18. **Operation Homefront** - Provides emergency financial assistance to military families and offers programs focused on housing and caregiver support. www.operationhomefront.org
19. **Fisher House Foundation** - Provides free housing for military families while a loved one is

receiving treatment at a VA medical center. www.fisherhouse.org

20. **Semper Fi & America's Fund** - A nonprofit organization offering financial assistance, transition support, and family aid to injured and ill service members.www.thefund.org
21. **The National Veterans Foundation** - Provides crisis management and information/referral services for veterans dealing with issues such as PTSD, homelessness, and unemployment. www.nvf.org
22. **AmericaServes** - A coordinated network of service providers helping veterans and their families with everything from housing to health care. www.americaserves.org
23. **Veterans Advantage** - A membership program providing veterans with discounts on travel, health services, and retail purchases. www.veteransadvantage.com
24. **Operation Gratitude** - Sends care packages to deployed military members, veterans, and first responders as a sign of appreciation for their service. www.operationgratitude.com
25. **Vets4Warriors** - A peer support network that provides confidential, one-on-one assistance for veterans by fellow veterans. www.vets4warriors.com
26. **Code of Support Foundation** - Provides personalized case coordination for veterans facing a range of challenges including financial, health, and legal issues. www.codeofsupport.org
27. **Veterans Green Jobs** - Offers job training and employment opportunities in environmentally

sustainable industries for veterans transitioning to civilian life. www.veteransgreenjobs.org

28. **Cohen Veterans Network** - A national network of mental health clinics offering free care to veterans and their families, focusing on PTSD and transition-related challenges. www.cohenveteransnetwork.org
29. **Homes for Our Troops** - A nonprofit that builds and donates specially adapted homes for severely injured post-9/11 veterans. www.hfotusa.org
30. **Veterans Resource Centers of America (VRCA)** - Provides a wide range of services, including housing, job training, and mental health care for veterans in need. www.vetsresource.org
31. **Building Homes for Heroes** - Builds and renovates homes for military veterans and their families to help them live independently. www.buildinghomesforheroes.org
32. **Team Red, White & Blue (Team RWB)** - Connects veterans with their communities through physical and social activities to improve health and well-being. www.teamrwb.org
33. **Got Your 6** - An initiative that works to ensure veterans are perceived as leaders and civic assets, bridging the gap between veterans and civilians. www.gotyour6.org
34. **Veterans Yoga Project** - A nonprofit organization that provides yoga classes for veterans dealing with PTSD, depression, and anxiety. www.veteransyogaproject.org
35. **National Coalition for Homeless Veterans (NCHV)** - Works to end homelessness among

veterans by providing housing, employment, and support services. www.nchv.org

36. **Iraq and Afghanistan Veterans of America (IAVA)** - A nonprofit that advocates for policy changes to benefit post-9/11 veterans and offers transition resources, mental health support, and career services. www.iava.org
37. **Veterans Service Organizations (VSOs)** - Provides advocacy, benefits assistance, and support for veterans through accredited organizations. www.va.gov/vso
38. **Boot Campaign** - Raises awareness of veteran issues and funds assistance programs for veterans through the sale of combat boots. www.bootcampaign.org
39. **Operation Stand Down** - Assists veterans who are homeless or at risk of homelessness by providing employment services, housing, and basic needs assistance. www.osdnashville.org
40. **Hope For The Warriors** - Offers transition support, education, and employment resources for post-9/11 veterans and their families. www.hopeforthewarriors.org
41. **Swords to Plowshares** - A veterans organization providing employment training, legal services, and housing assistance to veterans in need. www.swords-to-plowshares.org
42. **Operation Mend** - Provides free medical, surgical, and psychological treatment to severely injured veterans and their families. www.uclahealth.org/operationmend
43. **Military OneSource** - A Department of Defense resource providing confidential

support, counseling, and education services to veterans and their families. www.military-onesource.mil

44. **National Military Family Association (NMFA)** - Provides scholarships, transition support, and advocacy for military families, helping them adjust to life after service. www.militaryfamily.org
45. **American Corporate Partners (ACP)** - Provides one-on-one career mentorship to veterans transitioning into the corporate world. www.acp-usa.org
46. **Veterans of America Rehabilitation Program** - Offers services to assist veterans with mental and physical rehabilitation, career readiness, and reentry into civilian life. www.va.gov/rehabilitation
47. **American Red Cross – Services to Armed Forces** - Provides emergency services, financial assistance, and transition programs for veterans and military families. www.redcross.org/military
48. **Boulder Crest Foundation** - Offers healing retreats and wellness programs for combat veterans and their families to help them overcome PTSD and other combat-related trauma. www.bouldercrest.org
49. **Warrior-Scholar Project** - Offers academic boot camps to help enlisted veterans succeed in higher education at top universities across the U.S. www.warrior-scholar.org
50. **Combat Veterans Motorcycle Association (CVMA)** - A national organization of veterans who support fellow veterans through motorcycle

rides, charity events, and community outreach. www.combatvet.us

51. **Operation FINALLY HOME** - Provides mortgage-free homes to wounded, ill, and injured veterans, as well as widows of fallen soldiers. www.operationfinallyhome.org
52. **Veteran Tickets Foundation (Vet Tix)** - Provides free event tickets to veterans and their families, helping them connect with their communities through sports, concerts, and other events. www.vettix.org
53. **Paralyzed Veterans of America (PVA)** - A nonprofit organization that provides support for veterans suffering from spinal cord injuries or related diseases, including benefits, legal assistance, and advocacy. www.pva.org
54. **Adaptive Training Foundation** - Offers physical rehabilitation services and fitness training for injured veterans to help them regain strength and confidence. www.adaptivetraining-foundation.org
55. **Warrior Canine Connection** - Trains service dogs to assist veterans with PTSD and mobility impairments, improving their quality of life. www.warriorcanineconnection.org
56. **Operation Second Chance** - Provides financial assistance, peer support, and morale-boosting retreats for wounded veterans and their families. www.operationsecondchance.org
57. **Blue Star Families** - A nonprofit organization offering resources and support to military families, including career programs, benefits, and educational tools. www.bluestarfam.org

58. **Lone Survivor Foundation** - Provides mental health services and retreat programs for wounded service members, veterans, and their families. www.lonesurvivorfoundation.org
59. **Veteran Women Igniting the Spirit of Entrepreneurship (V-WISE)** - Offers entrepreneurship and small business training programs specifically for female veterans and military spouses. www.ivmf.syracuse.edu/vwise
60. **Hope for the Warriors** - Provides transition services, wellness programs, and peer support for veterans and military families, helping them adjust to civilian life. www.hopeforthe-warriors.org
61. **Save A Warrior** - Provides veterans with mental health services focused on treating PTSD and suicidal tendencies through an innovative warrior-led program. www.saveawarrior.org
62. **Military Veterans in Journalism (MVJ)** - Provides training, job placement, and networking opportunities for veterans seeking careers in journalism and media. www.mvj.network
63. **Headstrong Project** - Offers cost-free mental health treatment to veterans struggling with PTSD, depression, and anxiety. www.getheadstrong.org
64. **VetFran** - Provides veterans with access to franchise ownership opportunities, helping them start and grow businesses. www.vetfran.org
65. **Veterans for Common Sense (VCS)** - Advocates for policy changes and provides resources that improve the health and well-being of veterans. www.veteransforcommonsense.org

66. **The Honor Foundation** - Helps Special Operations veterans transition from military service into the private sector with career readiness programs and mentorship. www.honor.org
67. **Got Your Six Support Dogs** - Trains service dogs to help veterans with PTSD, providing life-saving support and companionship. www.gotyoursix-supportdogs.org
68. **Operation Supply Drop** - Supports veterans and active duty service members through gaming programs, helping them build connections and improve mental health. www.operationsupplydrop.org
69. **Dog Tag Bakery** - Provides entrepreneurship and business management training for veterans and their spouses through a hands-on bakery internship. www.dogtaginc.org
70. **America's Warrior Partnership** - A national organization that connects veterans with resources in their communities, focusing on housing, health care, and employment. www.americaswarriorpartnership.org
71. **Pets for Vets** - Matches veterans with shelter dogs trained to provide companionship and emotional support. www.petsforvets.com
72. **Veterans Justice Outreach (VJO)** - A VA program aimed at helping veterans involved with the justice system get treatment for mental health issues, rather than serving jail time. www.va.gov/homeless/vjo.asp
73. **Adaptive Adventures** - Provides outdoor sports and recreational opportunities for veterans with

physical disabilities to enhance their health and quality of life. www.adaptiveadventures.org

74. **National Veterans Legal Services Program (NVLSP)** - Offers free legal services to veterans seeking disability benefits and assists with VA claims and appeals. www.nvlsp.org
75. **Veterans Law Project** - Provides free legal representation and counseling to veterans and their families, focusing on benefits claims and appeals. www.veteranslawproject.org
76. **Military Hire** - A job board that connects veterans and transitioning service members with employers seeking their skills and experience. www.militaryhire.com
77. **Veterans in Piping (VIP)** - Offers training programs that teach veterans plumbing and welding skills, leading to certification and job placement. www.uavip.org
78. **Operation Healing Forces (OHF)** - Provides Special Operations veterans and their spouses with therapeutic retreats and ongoing support to promote mental and emotional recovery. www.operationhealingforces.org
79. **American Veterans (AMVETS)** - A nonprofit that provides advocacy, career services, and assistance with VA claims for all veterans, particularly those with disabilities. www.amvets.org
80. **Semper K9 Assistance Dogs** - Trains service dogs to assist disabled veterans, including those with PTSD and mobility challenges, at no cost. www.semperk9.org

81. **Sierra Club Military Outdoors** - Provides veterans with outdoor adventures such as hiking, camping, and rafting to promote healing and well-being.www.sierraclub.org/military-outdoors
82. **VA Adaptive Sports Programs** - Offers adaptive sports and recreational activities for veterans with disabilities, including skiing, cycling, and water sports. www.va.gov/adaptivesports
83. **Veterans Education Success (VES)** - Provides veterans with free legal services to help them protect their GI Bill benefits and pursue higher education. www.veteranseducationsuccess.org
84. **Warrior Rising** - Provides veterans with entrepreneurial resources and mentorship to help them start and grow businesses. www.warriorrising.org
85. **Military Child Education Coalition (MCEC)** - Supports the educational needs of military children by providing resources for families and schools. www.militarychild.org
86. **Give an Hour** - Provides free mental health care to veterans and their families through a national network of volunteer providers. www.giveanhour.org
87. **Student Veterans of America (SVA)** - A nonprofit organization that connects student veterans with resources to help them succeed in higher education and their careers. www.studentveterans.org
88. **Valor Service Dogs** - Trains and provides service dogs to veterans who suffer from PTSD, traumatic brain injuries, and mobility impairments. www.valorservicedogs.org

89. **Operation ReEngage** - Offers job placement and career coaching services for veterans transitioning into the workforce. www.operationreengage.org
90. **RallyPoint** - An online social network for veterans, active duty service members, and military spouses to connect, share resources, and find job opportunities. www.rallypoint.com
91. **Veteran Ventures Capital** - A venture capital fund that invests in veteran-owned businesses, providing them with capital and resources to grow. www.veteranventures.us
92. **Bunker Labs** - A national network of veteran entrepreneurs dedicated to helping veterans start and grow their own businesses. www.bunkerlabs.org
93. **Veteran Artist Program (VAP)** - Provides veterans with opportunities to pursue careers in the arts, including film, photography, music, and visual arts. www.veteranartistprogram.org
94. **Hero's Wish Program** - Grants wishes to terminally ill veterans, providing them with memorable experiences with their families. www.militarywarriors.org/heroswish
95. **Purple Heart Foundation** - Provides financial support and assistance programs for veterans who have been wounded in combat. www.purpleheartfoundation.org
96. **Combat Stress Recovery Program (CSRP)** - Offers mental health services for veterans who have experienced trauma, including PTSD, combat stress, and anxiety. www.

woundedwarriorproject.org/programs/combat-stress-recovery

97. **Homes on the Homefront** - Provides rent-free homes to veterans and their families, helping them transition into homeownership. www.operationhomefront.org/homesonthehomefront
98. **Honor Flight Network** - Honors veterans by transporting them to Washington, D.C., to visit the memorials dedicated to their service. www.honorflight.org
99. **American Gold Star Mothers** - Provides support to mothers who have lost a child in military service, offering community and volunteer opportunities. www.goldstarmoms.com
100. **Veterans Affordable Housing Program (VAHP)** - Provides affordable housing options and home repair services for veterans in need.www.vahousingprogram.org

VA CLAIMS PROCESS GUIDE

"Veterans are the backbone of our nation, and we have a solemn duty to ensure that they receive the benefits they've earned. If you served, you have earned these rights — don't wait to claim them."

— President Joe Biden

Step-by-Step VA Claims Process Guide

Navigating the VA claims process can be daunting, but with the right information and approach, veterans can access the benefits they've earned. This step-by-step guide simplifies the process:

Step 1: Determine Eligibility

Before applying for any benefits, it's important to know if you're eligible. Veterans must have served in active duty and been discharged under conditions other than dishonorable to qualify for most VA benefits.

- **Check VA eligibility criteria:** www.va.gov/health-care/eligibility/

Step 2: Gather Necessary Documents

Ensure you have all required documents ready, including:

- **DD Form 214** (discharge papers)
- **Medical records** (both military and civilian)
- **Dependency records** (for family benefits)
- **Service treatment records** (proof of injuries, if filing for disability)

Step 3: Choose a VA Representative

Veterans can file claims themselves, but working with an accredited Veterans Service Officer (VSO) can make the process easier. They offer free assistance in preparing and submitting claims.

- **Find a VSO near you:** www.va.gov/vso/

Step 4: Submit Your Application

Veterans can submit their claims online, by mail, or in person at a VA office.

- **Online:** Use the eBenefits portal at www.ebenefits.va.gov
- **By mail:** Download and mail VA Form 21-526EZ to your regional office.
- **In person:** Visit your nearest VA office for assistance.

Step 5: Attend VA Medical Examinations

Once your claim is submitted, the VA may schedule a Compensation & Pension (C&P) examination. This is essential for determining the level of disability benefits.

- **Prepare for your C&P exam:** Bring all medical records, prescriptions, and a list of symptoms related to your claim.

Step 6: Track Your Claim Status

You can track the status of your claim through the eBenefits portal or by contacting the VA directly.

- **eBenefits:** www.ebenefits.va.gov
- **VA Hotline:** 1-800-827-1000

Step 7: Receive a Decision

Once the VA processes your claim, you will receive a decision letter. This will outline whether your claim was approved, the percentage of disability granted, and your compensation amount.

Step 8: Appeal, If Necessary

If you disagree with the VA's decision, you have the right to appeal. You must file a Notice of Disagreement (NOD) within one year of receiving your decision letter.

- **How to appeal a decision:** www.va.gov/decision-reviews/

Step 9: Manage Your Benefits

Once approved, veterans will receive their benefits on a monthly basis. Use the VA.gov portal to manage your benefits, update personal information, and apply for additional services as needed.

. . .

VETERANS BENEFITS CHEAT **Sheet**

This quick-reference guide provides a high-level overview of the most common VA benefits available to veterans and their families, helping you navigate the essential programs.

1. Disability Compensation

- **What it is:** Tax-free monetary benefit for veterans with disabilities resulting from service-connected injuries or illnesses.
- **How to apply:** File VA Form 21-526EZ via eBenefits
- **Eligibility:** Service-related disabilities, injuries, or diseases.
- **Payment amount:** Based on the degree of disability (10% - 100%).

2. VA Healthcare

- **What it is:** Comprehensive healthcare services, including primary care, specialty care, mental health, and more.
- **How to apply:** Enroll at VA Healthcare
- **Eligibility:** Veterans who served on active duty and were discharged under conditions other than dishonorable.

3. GI Bill

- **What it is:** Educational benefits covering tuition, books, housing, and other education-related expenses.
- **How to apply:** Use the GI Bill application
- **Eligibility:** Based on time served, usually for veterans with at least 90 days of active duty post-9/11.
- **Tip:** Don't forget to check eligibility for **Yellow Ribbon Program** for additional funding.

4. VA Home Loan Program

- **What it is:** Provides guaranteed loans to help veterans buy, build, or improve a home.
- **How to apply:** Work with approved lenders via the VA Home Loan site
- **Eligibility:** Must have satisfactory credit, sufficient income, and a valid Certificate of Eligibility (COE).

5. Vocational Rehabilitation and Employment (VR&E)

- **What it is:** Assistance with job training, employment accommodations, resume development, and more.
- **How to apply:** Visit the VR&E page
- **Eligibility:** Veterans with a service-connected disability of 10% or more.

6. Life Insurance (SGLI, VGLI)

- **What it is:** Life insurance for veterans transitioning from the military, with affordable premiums.
- **How to apply:** SGLI to VGLI conversion
- **Eligibility:** Veterans who had SGLI during active duty can convert to VGLI within 1 year and 120 days of separation.

7. Survivor Benefits

- **What it is:** Financial benefits for surviving spouses, children, or parents of deceased veterans.
- **How to apply:** Survivors can file claims for Dependency and Indemnity Compensation (DIC) through the VA benefits page
- **Eligibility:** Veterans who died from service-connected disabilities or were totally disabled at the time of death.

8. Burial and Memorial Benefits

- **What it is:** Offers burial in VA national cemeteries, grave markers, and a burial allowance.
- **How to apply:** Pre-need eligibility can be determined through the VA burial benefits page
- **Eligibility:** Veterans who were discharged under conditions other than dishonorable.

9. Mental Health Services

- **What it is:** Provides counseling, therapy, and treatment for veterans dealing with PTSD, depression, anxiety, and more.
- **How to apply:** Access services through the VA Mental Health portal
- **Eligibility:** Any veteran enrolled in VA healthcare is eligible for mental health services.

10. Caregiver Support Program

- **What it is:** Provides resources, training, and financial assistance to caregivers of veterans.
- **How to apply:** Apply for caregiver benefits via the VA Caregiver Support Program
- **Eligibility:** Available to caregivers of veterans who need personal care services due to serious injuries sustained in the line of duty.

1. Step-by-Step VA Claims Process Guide

- Provide a detailed breakdown of how veterans can navigate the VA claims process for benefits, including how to file for disability, healthcare, education, and more. Include common pitfalls, timelines, and how to seek assistance.
- **Value:** Many veterans struggle with understanding the complexities of the VA system. A clear, step-by-step guide could ease this process.

2. Financial Planning and Budgeting Tools

- Create or link to financial planning tools that help veterans budget for civilian life, plan for retirement, or manage benefits like pensions and GI Bill funds. Include a worksheet or online tool veterans can use to estimate their financial needs.
- **Value:** Transitioning from military to civilian life often brings a big change in finances. Tools to plan and budget effectively can help veterans manage this shift smoothly.

3. Mental Health Self-Assessment and Resource Guide

- Include a mental health self-assessment tool that veterans can use to evaluate their emotional well-being, followed by a list of local and national mental health resources they can access.
- **Value:** Veterans may face mental health challenges like PTSD, anxiety, or depression after service. A self-assessment tool can help them recognize early signs and access support sooner.

4. Resume Templates and Job Application Tips

- Offer downloadable resume templates specifically tailored to veterans, showing how to translate military skills into civilian job qualifications. Provide job application tips, such as how to network effectively and prepare for interviews.
- **Value:** Writing resumes and understanding how to apply for civilian jobs can be challenging.

Practical examples and templates could help veterans make a strong first impression.

5. Veterans Community and Networking Opportunities

- Include information about local and online veteran support groups, job fairs, and veteran networking events. List professional organizations that welcome veterans, including those related to entrepreneurship, specific industries, or veteran-owned business communities.
- **Value:** Transitioning veterans may feel isolated. Connecting with a community of peers through events or support networks can provide social and professional connections that boost confidence and career prospects.

6. Scholarship and Educational Grant Opportunities

- Provide a list of scholarships and educational grants exclusively available to veterans and their families. Break it down by level (undergraduate, graduate, vocational training) and include resources for non-traditional education options.
- **Value:** Beyond the GI Bill, veterans often qualify for other scholarships and grants. Highlighting these opportunities could encourage veterans to further their education or pursue new career paths.

7. Housing Assistance Programs

- Offer a guide on housing options, including VA loans, rental assistance programs, and homeownership programs for veterans. Explain how veterans can take advantage of these benefits and any eligibility criteria they need to meet.
- **Value:** Finding affordable housing can be a challenge for veterans, especially those with disabilities or who have recently left service. A clear guide to housing options can reduce stress during the transition.

8. Veterans Benefits Cheat Sheet

- Create a quick-reference "cheat sheet" that summarizes key veteran benefits (like healthcare, disability, housing, education), including eligibility requirements and how to apply. Keep it simple and user-friendly.
- **Value:** Veterans often feel overwhelmed by the sheer number of benefits and programs available. A concise cheat sheet could provide easy access to the most important information.

9. Veteran Entrepreneurship Resources

- Include links and contact information for organizations that provide mentorship, startup funding, and business planning services for veterans interested in entrepreneurship. Add case studies of successful veteran-owned businesses for inspiration.

- **Value:** Many veterans are interested in starting their own businesses. Offering resources for entrepreneurship could be a valuable addition to the guide.

These suggestions will provide practical, hands-on resources that can help veterans take action during their transition, from personal well-being and financial stability to professional development and community support.

Scholarship and Educational Grant Opportunities

Veterans and their families have access to a variety of scholarships and educational grants to further their education and career development. Here are key resources for obtaining financial aid.

1. GI Bill®

- **What it is:** The U.S. Department of Veterans Affairs offers education benefits that help cover tuition, housing, books, and other related expenses.
- **Who it helps:** Veterans, service members, and dependents.
- **Link:** www.benefits.va.gov/gibill

2. Yellow Ribbon Program

- **What it is:** An extension of the GI Bill that covers additional tuition costs at private schools and out-of-state public universities.

- **Who it helps:** Veterans attending schools that participate in the Yellow Ribbon Program.
- **Link:** www.benefits.va.gov/gibill/yellow_ribbon.asp

3. Tillman Scholar Program (Pat Tillman Foundation)

- **What it is:** Provides educational scholarships to military veterans and their spouses who show leadership potential and a desire to impact their communities.
- **Who it helps:** Veterans and their spouses pursuing higher education.
- **Link:** www.ptf.org

4. AMVETS National Scholarship Program

- **What it is:** Offers scholarships for veterans, active-duty service members, and their children.
- **Who it helps:** Veterans and their families pursuing undergraduate or graduate degrees.
- **Link:** www.amvets.org/scholarships

5. Veterans of Foreign Wars (VFW) Scholarships

- **What it is:** Provides a variety of scholarships and awards for veterans and their dependents, including the "Sport Clips Help A Hero Scholarship."
- **Who it helps:** Veterans attending trade schools or pursuing higher education.
- **Link:** www.vfw.org/scholarship

6. AFCEA Educational Foundation Veterans Scholarships

- **What it is:** Provides merit-based scholarships to veterans pursuing degrees in STEM fields (Science, Technology, Engineering, Math).
- **Who it helps:** Veterans pursuing undergraduate or graduate degrees in STEM.
- **Link:** www.afcea.org/site/foundation/scholarships

7. Marine Gunnery Sergeant John David Fry Scholarship

- **What it is:** Provides education benefits to children and surviving spouses of service members who died in the line of duty after September 10, 2001.
- **Who it helps:** Children and spouses of deceased veterans.
- **Link:** www.benefits.va.gov/gibill/fry_scholarship.asp

8. Google Student Veterans of America (SVA) Scholarship

- **What it is:** Provides scholarships to student veterans pursuing degrees in computer science or related fields.
- **Who it helps:** Veterans in STEM or computer science-related programs.
- **Link:** www.studentveterans.org/programs/scholarships

9. Army Women's Foundation Legacy Scholarship

- **What it is:** Provides scholarships for women who have served or are serving in the U.S. Army and their descendants.
- **Who it helps:** Female veterans and their children pursuing higher education.
- **Link:** www.awfdn.org

10. Navy-Marine Corps Relief Society Education Assistance

- **What it is:** Offers interest-free loans and grants for education to children and spouses of active-duty, retired, or deceased Navy and Marine Corps service members.
- **Who it helps:** Children and spouses of Navy and Marine Corps service members.
- **Link:** www.nmcrs.org

11. Iraq and Afghanistan Service Grant

- **What it is:** A grant for students whose parent or guardian died as a result of military service in Iraq or Afghanistan after the events of 9/11.
- **Who it helps:** Dependents of veterans who died in Iraq or Afghanistan.
- **Link:** www.studentaid.gov/understand-aid/types/grants/iraq-afghanistan-service

12. Wings Over America Scholarship Foundation

- **What it is:** Provides college scholarships to dependent children and spouses of Navy personnel.
- **Who it helps:** Children and spouses of Navy service members.
- **Link:** www.wingsoveramerica.us

13. **Anchor Scholarship Foundation**

- **What it is:** Provides financial aid to qualified dependents of U.S. Navy Surface Warfare personnel.
- **Who it helps:** Children and spouses of U.S. Navy Surface Warfare personnel.
- **Link:** www.anchorscholarship.com

14. **Raytheon Patriot Scholarship**

- **What it is:** A scholarship offered to veterans pursuing undergraduate degrees in STEM fields.
- **Who it helps:** Veterans studying STEM fields.
- **Link:** www.studentveterans.org/programs/scholarships

15. **Military Order of the Purple Heart Scholarship**

- **What it is:** Provides scholarships for Purple Heart recipients, their spouses, and children.
- **Who it helps:** Purple Heart recipients and their dependents pursuing higher education.
- **Link:** www.purpleheart.org/scholarship-program

16. **Folds of Honor Scholarship**

- **What it is:** Offers scholarships to the spouses and children of fallen or disabled service members.
- **Who it helps:** Spouses and children of fallen or disabled service members.
- **Link:** www.foldsofhonor.org/scholarships

17. Troops to Teachers Program

- **What it is:** A grant program that provides financial assistance for veterans seeking to become certified teachers.
- **Who it helps:** Veterans transitioning to careers in education.
- **Link:** www.proudtoserveagain.com

18. Military Spouse Career Advancement Account (MyCAA)

- **What it is:** Offers financial assistance for education and training for military spouses pursuing portable careers.
- **Who it helps:** Spouses of active-duty service members seeking certifications or degrees.
- **Link:** www.militaryonesource.mil/mycaa

19. Student Veterans of America (SVA) National Scholarship Program

- **What it is:** Provides scholarships for student veterans to pursue post-secondary education.
- **Who it helps:** Veterans enrolled in college or university programs.

- **Link:** www.studentveterans.org/programs/scholarships

20. AFSA Scholarship Program (Air Force Sergeants Association)

- **What it is:** Offers scholarships to Air Force personnel and their families, including dependent children and spouses.
- **Who it helps:** Air Force members, spouses, and dependents.
- **Link:** www.hqafsa.org/scholarships

21. The Army Scholarship Foundation

- **What it is:** Provides scholarships to Army soldiers, veterans, and their family members for undergraduate study.
- **Who it helps:** Active duty and retired Army personnel, veterans, and their dependents.
- **Link:** www.armyscholarshipfoundation.org

22. Veterans United Foundation Scholarship

- **What it is:** Offers scholarships to veterans, military members, and their families pursuing higher education.
- **Who it helps:** Veterans, service members, and dependents pursuing undergraduate or graduate degrees.
- **Link:** www.enhancelives.com/scholarships

23. American Legion Auxiliary Scholarship

- **What it is:** Provides several scholarships to military veterans and their family members, including the Children of Warriors National Presidents' Scholarship.
- **Who it helps:** Veterans, military service members, and their families.
- **Link:** www.alaforveterans.org/scholarships

24. No Greater Sacrifice Foundation Scholarship

- **What it is:** Provides scholarships to the children of fallen and severely wounded service members to attend college or vocational schools.
- **Who it helps:** Children of fallen or severely wounded service members.
- **Link:** www.nogreatersacrifice.org

25. Marine Corps Scholarship Foundation

- **What it is:** Offers need-based scholarships to children of active duty or veteran Marines for higher education.
- **Who it helps:** Children of Marines pursuing undergraduate or vocational education.
- **Link:** www.mcsf.org

26. Freedom Alliance Scholarship Fund

- **What it is:** Provides scholarships to children of service members who were killed or permanently disabled in military service.
- **Who it helps:** Children of fallen or disabled veterans.

- **Link:** www.freedomalliance.org

27. Special Operations Warrior Foundation Scholarship

- **What it is:** Offers college scholarships to children of Special Operations service members who have died in the line of duty.
- **Who it helps:** Children of fallen Special Operations forces.
- **Link:** www.specialops.org

28. American Patriot Freedom Scholarship

- **What it is:** Provides scholarships to military families, particularly for high school seniors planning to attend college.
- **Who it helps:** Children of active duty and veteran service members.
- **Link:** www.homefrontamerica.org/scholarships

29. Children of Fallen Patriots Foundation

- **What it is:** Provides scholarships and educational counseling to children who have lost a parent in military service.
- **Who it helps:** Children of fallen service members.
- **Link:** www.fallenpatriots.org

30. Veterans of Foreign Wars Sport Clips Help A Hero Scholarship

- **What it is:** Provides scholarships to veterans and service members for use at post-secondary institutions or trade schools.
- **Who it helps:** Veterans and active duty military members.
- **Link:** www.vfw.org/scholarship

31. Wings for Warriors Scholarship

- **What it is:** Offers scholarships to veterans who are pursuing higher education in healthcare-related fields.
- **Who it helps:** Veterans entering healthcare fields.
- **Link:** www.wingsforwarriors.org/scholarship

32. USAA Scholarship Program

- **What it is:** Provides scholarships to military members, veterans, and their families through various military-affiliated organizations.
- **Who it helps:** Veterans, service members, and their dependents.
- **Link:** www.usaa.com/inet/pages/scholarships

33. Elks National Foundation Veterans Scholarship

- **What it is:** Provides scholarships to veterans and their family members, including children and grandchildren.
- **Who it helps:** Veterans, military members, and their families pursuing higher education.
- **Link:** www.elks.org/vets

34. Air Force Aid Society (AFAS) General Henry H. Arnold Education Grant Program

- **What it is:** Offers education grants to Air Force members and their dependents.
- **Who it helps:** Active duty, retired, or deceased Air Force personnel and their families.
- **Link:** www.afas.org/education-support

35. ThanksUSA Scholarship Program

- **What it is:** Provides need-based scholarships to military spouses and children for college and vocational training.
- **Who it helps:** Spouses and children of active-duty military personnel.
- **Link:** www.thanksusa.org

VETERANS BENEFITS CHEAT SHEET

This quick-reference guide provides a high-level overview of the most common VA benefits available to veterans and their families, helping you navigate the essential programs.

1. Disability Compensation

- **What it is:** Tax-free monetary benefit for veterans with disabilities resulting from service-connected injuries or illnesses.
- **How to apply:** File VA Form 21-526EZ via eBenefits
- **Eligibility:** Service-related disabilities, injuries, or diseases.
- **Payment amount:** Based on the degree of disability (10% - 100%).

2. VA Healthcare

- **What it is:** Comprehensive healthcare services, including primary care, specialty care, mental health, and more.
- **How to apply:** Enroll at VA Healthcare
- **Eligibility:** Veterans who served on active duty and were discharged under conditions other than dishonorable.

3. GI Bill

- **What it is:** Educational benefits covering tuition, books, housing, and other education-related expenses.
- **How to apply:** Use the GI Bill application
- **Eligibility:** Based on time served, usually for veterans with at least 90 days of active duty post-9/11.
- **Tip:** Don't forget to check eligibility for **Yellow Ribbon Program** for additional funding.

4. VA Home Loan Program

- **What it is:** Provides guaranteed loans to help veterans buy, build, or improve a home.
- **How to apply:** Work with approved lenders via the VA Home Loan site
- **Eligibility:** Must have satisfactory credit, sufficient income, and a valid Certificate of Eligibility (COE).

5. Vocational Rehabilitation and Employment (VR&E)

- **What it is:** Assistance with job training, employment accommodations, resume development, and more.
- **How to apply:** Visit the VR&E page
- **Eligibility:** Veterans with a service-connected disability of 10% or more.

6. Life Insurance (SGLI, VGLI)

- **What it is:** Life insurance for veterans transitioning from the military, with affordable premiums.
- **How to apply:** SGLI to VGLI conversion
- **Eligibility:** Veterans who had SGLI during active duty can convert to VGLI within 1 year and 120 days of separation.

7. Survivor Benefits

- **What it is:** Financial benefits for surviving spouses, children, or parents of deceased veterans.
- **How to apply:** Survivors can file claims for Dependency and Indemnity Compensation (DIC) through the VA benefits page
- **Eligibility:** Veterans who died from service-connected disabilities or were totally disabled at the time of death.

8. Burial and Memorial Benefits

- **What it is:** Offers burial in VA national cemeteries, grave markers, and a burial allowance.
- **How to apply:** Pre-need eligibility can be determined through the VA burial benefits page
- **Eligibility:** Veterans who were discharged under conditions other than dishonorable.

9. Mental Health Services

- **What it is:** Provides counseling, therapy, and treatment for veterans dealing with PTSD, depression, anxiety, and more.
- **How to apply:** Access services through the VA Mental Health portal
- **Eligibility:** Any veteran enrolled in VA healthcare is eligible for mental health services.

10. Caregiver Support Program

- **What it is:** Provides resources, training, and financial assistance to caregivers of veterans.
- **How to apply:** Apply for caregiver benefits via the VA Caregiver Support Program
- **Eligibility:** Available to caregivers of veterans who need personal care services due to serious injuries sustained in the line of duty.

1. Step-by-Step VA Claims Process Guide

- Provide a detailed breakdown of how veterans can navigate the VA claims process for benefits, including how to file for disability, healthcare, education, and more. Include common pitfalls, timelines, and how to seek assistance.
- **Value:** Many veterans struggle with understanding the complexities of the VA system. A clear, step-by-step guide could ease this process.

2. Financial Planning and Budgeting Tools

- Create or link to financial planning tools that help veterans budget for civilian life, plan for retirement, or manage benefits like pensions and GI Bill funds. Include a worksheet or online tool veterans can use to estimate their financial needs.
- **Value:** Transitioning from military to civilian life often brings a big change in finances. Tools to plan and budget effectively can help veterans manage this shift smoothly.

3. Mental Health Self-Assessment and Resource Guide

- Include a mental health self-assessment tool that veterans can use to evaluate their emotional well-being, followed by a list of local and national mental health resources they can access.
- **Value:** Veterans may face mental health challenges like PTSD, anxiety, or depression after service. A self-assessment tool can help them recognize early signs and access support sooner.

4. Resume Templates and Job Application Tips

- Offer downloadable resume templates specifically tailored to veterans, showing how to translate military skills into civilian job qualifications. Provide job application tips, such as how to network effectively and prepare for interviews.
- **Value:** Writing resumes and understanding how to apply for civilian jobs can be challenging. Practical examples and templates could help veterans make a strong first impression.

5. Veterans Community and Networking Opportunities

- Include information about local and online veteran support groups, job fairs, and veteran networking events. List professional organizations that welcome veterans, including those related to entrepreneurship, specific industries, or veteran-owned business communities.
- **Value:** Transitioning veterans may feel isolated. Connecting with a community of peers through events or support networks can provide social and professional connections that boost confidence and career prospects.

6. Scholarship and Educational Grant Opportunities

- Provide a list of scholarships and educational grants exclusively available to veterans and

their families. Break it down by level (undergraduate, graduate, vocational training) and include resources for non-traditional education options.
- **Value:** Beyond the GI Bill, veterans often qualify for other scholarships and grants. Highlighting these opportunities could encourage veterans to further their education or pursue new career paths.

7. Housing Assistance Programs

- Offer a guide on housing options, including VA loans, rental assistance programs, and homeownership programs for veterans. Explain how veterans can take advantage of these benefits and any eligibility criteria they need to meet.
- **Value:** Finding affordable housing can be a challenge for veterans, especially those with disabilities or who have recently left service. A clear guide to housing options can reduce stress during the transition.

8. Veterans Benefits Cheat Sheet

- Create a quick-reference "cheat sheet" that summarizes key veteran benefits (like healthcare, disability, housing, education), including eligibility requirements and how to apply. Keep it simple and user-friendly.
- **Value:** Veterans often feel overwhelmed by the sheer number of benefits and programs

available. A concise cheat sheet could provide easy access to the most important information.

9. Veteran Entrepreneurship Resources

- Include links and contact information for organizations that provide mentorship, startup funding, and business planning services for veterans interested in entrepreneurship. Add case studies of successful veteran-owned businesses for inspiration.
- **Value:** Many veterans are interested in starting their own businesses. Offering resources for entrepreneurship could be a valuable addition to the guide.

SCHOLARSHIP AND GRANTS

Scholarship and Educational Grant Opportunities

Veterans and their families have access to a variety of scholarships and educational grants to further their education and career development. Here are key resources for obtaining financial aid.

1. GI Bill®

- **What it is:** The U.S. Department of Veterans Affairs offers education benefits that help cover tuition, housing, books, and other related expenses.
- **Who it helps:** Veterans, service members, and dependents.
- **Link:** www.benefits.va.gov/gibill

2. Yellow Ribbon Program

- **What it is:** An extension of the GI Bill that covers additional tuition costs at private schools and out-of-state public universities.
- **Who it helps:** Veterans attending schools that participate in the Yellow Ribbon Program.
- **Link:** www.benefits.va.gov/gibill/yellow_ribbon.asp

3. Tillman Scholar Program (Pat Tillman Foundation)

- **What it is:** Provides educational scholarships to military veterans and their spouses who show leadership potential and a desire to impact their communities.
- **Who it helps:** Veterans and their spouses pursuing higher education.
- **Link:** www.ptf.org

4. AMVETS National Scholarship Program

- **What it is:** Offers scholarships for veterans, active-duty service members, and their children.
- **Who it helps:** Veterans and their families pursuing undergraduate or graduate degrees.
- **Link:** www.amvets.org/scholarships

5. Veterans of Foreign Wars (VFW) Scholarships

- **What it is:** Provides a variety of scholarships and awards for veterans and their dependents, including the "Sport Clips Help A Hero Scholarship."

- **Who it helps:** Veterans attending trade schools or pursuing higher education.
- **Link:** www.vfw.org/scholarship

6. AFCEA Educational Foundation Veterans Scholarships

- **What it is:** Provides merit-based scholarships to veterans pursuing degrees in STEM fields (Science, Technology, Engineering, Math).
- **Who it helps:** Veterans pursuing undergraduate or graduate degrees in STEM.
- **Link:** www.afcea.org/site/foundation/scholarships

7. Marine Gunnery Sergeant John David Fry Scholarship

- **What it is:** Provides education benefits to children and surviving spouses of service members who died in the line of duty after September 10, 2001.
- **Who it helps:** Children and spouses of deceased veterans.
- **Link:** www.benefits.va.gov/gibill/fry_scholarship.asp

8. Google Student Veterans of America (SVA) Scholarship

- **What it is:** Provides scholarships to student veterans pursuing degrees in computer science or related fields.
- **Who it helps:** Veterans in STEM or computer science-related programs.
- **Link:** www.studentveterans.org/programs/scholarships

9. Army Women's Foundation Legacy Scholarship

- **What it is:** Provides scholarships for women who have served or are serving in the U.S. Army and their descendants.
- **Who it helps:** Female veterans and their children pursuing higher education.
- **Link:** www.awfdn.org

10. Navy-Marine Corps Relief Society Education Assistance

- **What it is:** Offers interest-free loans and grants for education to children and spouses of active-duty, retired, or deceased Navy and Marine Corps service members.
- **Who it helps:** Children and spouses of Navy and Marine Corps service members.
- **Link:** www.nmcrs.org

11. Iraq and Afghanistan Service Grant

- **What it is:** A grant for students whose parent or guardian died as a result of military service in Iraq or Afghanistan after the events of 9/11.

- **Who it helps:** Dependents of veterans who died in Iraq or Afghanistan.
- **Link:** www.studentaid.gov/understand-aid/types/grants/iraq-afghanistan-service

12. **Wings Over America Scholarship Foundation**

- **What it is:** Provides college scholarships to dependent children and spouses of Navy personnel.
- **Who it helps:** Children and spouses of Navy service members.
- **Link:** www.wingsoveramerica.us

13. **Anchor Scholarship Foundation**

- **What it is:** Provides financial aid to qualified dependents of U.S. Navy Surface Warfare personnel.
- **Who it helps:** Children and spouses of U.S. Navy Surface Warfare personnel.
- **Link:** www.anchorscholarship.com

14. **Raytheon Patriot Scholarship**

- **What it is:** A scholarship offered to veterans pursuing undergraduate degrees in STEM fields.
- **Who it helps:** Veterans studying STEM fields.
- **Link:** www.studentveterans.org/programs/scholarships

15. **Military Order of the Purple Heart Scholarship**

- **What it is:** Provides scholarships for Purple Heart recipients, their spouses, and children.
- **Who it helps:** Purple Heart recipients and their dependents pursuing higher education.
- **Link:** www.purpleheart.org/scholarship-program

16. Folds of Honor Scholarship

- **What it is:** Offers scholarships to the spouses and children of fallen or disabled service members.
- **Who it helps:** Spouses and children of fallen or disabled service members.
- **Link:** www.foldsofhonor.org/scholarships

17. Troops to Teachers Program

- **What it is:** A grant program that provides financial assistance for veterans seeking to become certified teachers.
- **Who it helps:** Veterans transitioning to careers in education.
- **Link:** www.proudtoserveagain.com

18. Military Spouse Career Advancement Account (MyCAA)

- **What it is:** Offers financial assistance for education and training for military spouses pursuing portable careers.
- **Who it helps:** Spouses of active-duty service members seeking certifications or degrees.
- **Link:** www.militaryonesource.mil/mycaa

19. Student Veterans of America (SVA) National Scholarship Program

- **What it is:** Provides scholarships for student veterans to pursue post-secondary education.
- **Who it helps:** Veterans enrolled in college or university programs.
- **Link:** www.studentveterans.org/programs/scholarships

20. AFSA Scholarship Program (Air Force Sergeants Association)

- **What it is:** Offers scholarships to Air Force personnel and their families, including dependent children and spouses.
- **Who it helps:** Air Force members, spouses, and dependents.
- **Link:** www.hqafsa.org/scholarships

21. The Army Scholarship Foundation

- **What it is:** Provides scholarships to Army soldiers, veterans, and their family members for undergraduate study.
- **Who it helps:** Active duty and retired Army personnel, veterans, and their dependents.
- **Link:** www.armyscholarshipfoundation.org

22. Veterans United Foundation Scholarship

- **What it is:** Offers scholarships to veterans, military members, and their families pursuing higher education.
- **Who it helps:** Veterans, service members, and dependents pursuing undergraduate or graduate degrees.
- **Link:** www.enhancelives.com/scholarships

23. American Legion Auxiliary Scholarship

- **What it is:** Provides several scholarships to military veterans and their family members, including the Children of Warriors National Presidents' Scholarship.
- **Who it helps:** Veterans, military service members, and their families.
- **Link:** www.alaforveterans.org/scholarships

24. No Greater Sacrifice Foundation Scholarship

- **What it is:** Provides scholarships to the children of fallen and severely wounded service members to attend college or vocational schools.
- **Who it helps:** Children of fallen or severely wounded service members.
- **Link:** www.nogreatersacrifice.org

25. Marine Corps Scholarship Foundation

- **What it is:** Offers need-based scholarships to children of active duty or veteran Marines for higher education.

- **Who it helps:** Children of Marines pursuing undergraduate or vocational education.
- **Link:** www.mcsf.org

26. Freedom Alliance Scholarship Fund

- **What it is:** Provides scholarships to children of service members who were killed or permanently disabled in military service.
- **Who it helps:** Children of fallen or disabled veterans.
- **Link:** www.freedomalliance.org

27. Special Operations Warrior Foundation Scholarship

- **What it is:** Offers college scholarships to children of Special Operations service members who have died in the line of duty.
- **Who it helps:** Children of fallen Special Operations forces.
- **Link:** www.specialops.org

28. American Patriot Freedom Scholarship

- **What it is:** Provides scholarships to military families, particularly for high school seniors planning to attend college.
- **Who it helps:** Children of active duty and veteran service members.
- **Link:** www.homefrontamerica.org/scholarships

29. Children of Fallen Patriots Foundation

- **What it is:** Provides scholarships and educational counseling to children who have lost a parent in military service.
- **Who it helps:** Children of fallen service members.
- **Link:** www.fallenpatriots.org

30. Veterans of Foreign Wars Sport Clips Help A Hero Scholarship

- **What it is:** Provides scholarships to veterans and service members for use at post-secondary institutions or trade schools.
- **Who it helps:** Veterans and active duty military members.
- **Link:** www.vfw.org/scholarship

31. Wings for Warriors Scholarship

- **What it is:** Offers scholarships to veterans who are pursuing higher education in healthcare-related fields.
- **Who it helps:** Veterans entering healthcare fields.
- **Link:** www.wingsforwarriors.org/scholarship

32. USAA Scholarship Program

- **What it is:** Provides scholarships to military members, veterans, and their families through various military-affiliated organizations.
- **Who it helps:** Veterans, service members, and their dependents.

- **Link:** www.usaa.com/inet/pages/scholarships

33. Elks National Foundation Veterans Scholarship

- **What it is:** Provides scholarships to veterans and their family members, including children and grandchildren.
- **Who it helps:** Veterans, military members, and their families pursuing higher education.
- **Link:** www.elks.org/vets

34. Air Force Aid Society (AFAS) General Henry H. Arnold Education Grant Program

- **What it is:** Offers education grants to Air Force members and their dependents.
- **Who it helps:** Active duty, retired, or deceased Air Force personnel and their families.
- **Link:** www.afas.org/education-support

35. ThanksUSA Scholarship Program

- **What it is:** Provides need-based scholarships to military spouses and children for college and vocational training.
- **Who it helps:** Spouses and children of active-duty military personnel.
- **Link:** www.thanksusa.org

ALSO BY MS. VETERAN AFFAIRS

Life After Military Resources:

Life After Military Workbook

Life After Military Book

Life After Military Workbook - Spouse Version

Life After Military Book - Spouse Version

Veteran Empowerment Affirmations Series:

Self-Worth & Identity Affirmations, (Vol. 1)

Resilience & Adaptability Affirmations, (Vol. 2)

Career Transition & Professional Growth Affirmations, (Vol. 3)

Spousal Career & Personal Growth Affirmations, (Vol. 4)

Mental & Emotional Wellness Affirmations, (Vol. 5)

Available at Bookstores and MsVeteranAffairs.com

Made in United States
Orlando, FL
04 November 2026

90207807R00125